Computer Networks

Ata Elahi • Alex Cushman

Computer Networks

Data Communications, Internet and Security

 Springer

Ata Elahi
Southern Connecticut State University
New Haven, CT, USA

Alex Cushman
Southern Connecticut State University
New Haven, CT, USA

ISBN 978-3-031-42017-7   ISBN 978-3-031-42018-4   (eBook)
https://doi.org/10.1007/978-3-031-42018-4

This Springer imprint is published by the registered company Springer Nature Switzerland AG
The registered company address is: Gewerbestrasse 11, 6330 Cham, Switzerland

Paper in this product is recyclable.

*In the memory of my parents and
father in-law.*

Ata Elahi

Preface

This book is the result of teaching various networking courses at Southern Connecticut State University since 1987. The textbook covers: Digital Communications, Local Area Networks (LANs), Internet Protocols, Voice Over IP (VOIP), Wireless Local Area Networks (WLANs), Low Power Wireless Technologies (ZigBee, 6LowPAN, and LoRa), Cryptography, and Network Security. The beta version of the textbook was tested for use as a text reference for the undergraduate courses in networking at Southern Connecticut State University.

Intended Audience

This book is written primarily for students majoring in information technology, computer science, engineering technologies, or computer information systems, as well as any person interested in learning the concepts of computer networking. The book covers a broad range of topics and it can be used in higher level courses.

Organization

The materials of this book are presented using a practical approach, as opposed to using theory only, and thus no special background is required to understand the topics.

Chapter 1 – *Introduction to Communications Networks*: Network models, Network topologies, Types of Networks, the OSI model, the TCP/IP Model, Standard Organizations, and Communication Protocols

Chapter 2 – *Data Communications*: Analog Signals, Digital Signals, Binary Numbers, ASCII code, Full-duplex, Half-duplex, Serial and Parallel Transmission, Baseband Transmission, Broadband Transmission, Error Detection Methods (Parity

Check, Block Check Character (BCC), One's Complement of the Sum, and Cyclic Redundancy Check (CRC))

Chapter 3 – *Communications Channels and Media*: UTP, Fiber Optic Cabling, Wireless, Channel Bandwidth, Latency, Synchronous Optical Network (SONET), SONET Signal Rates, SONET Frame Format

Chapter 4 – *Multiplexer and Switching Concepts*: Types of Multiplexers (TDM, FDM, CDM, and WDM), Digitizing Voice, T1 Links, Circuit Switching, Packet Switching, and Virtual Circuits

Chapter 5 – *Error and Flow Control*: The Data Link Layer, Frame Transmission Methods, Flow Control (Stop and Wait ARQ, Continuous ARQ, and Sliding Window), and IEEE 802 Standard Committee

Chapter 6 – *Modulation Methods, Cable Modems, and FTTH*: Modem Operation, Modulation Methods (ASK, FSK, PSK, and QAM), Cable Modems, and Fiber to The Home (FTTH)

Chapter 7 – *Ethernet Technologies*: Ethernet, Fast Ethernet, Gigabit Ethernet, 10 Gigabit Ethernet, and Ethernet Access Methods

Chapter 8 – *LAN Interconnection Devices*: Repeaters, Bridges, Switches, Spanning Tree Protocol (STP), Layer 3 Switches, Virtual LAN (VLAN), VLAN Operation, Routers, and Gateways

Chapter 9 – *Internet Protocols Part I*: The Internet Architecture Board (IAB), TCP/IP Reference Model, TCP/IP Application Level, Transport Level Protocols (UDP and TCP), Internet Level Protocols (IP), IPv_4 Addressing, Classless IPV4 Addresses, ARP (Address Resolution Protocol), Internet Protocol version 6 (IPv6), and IPV6 Address Format

Chapter 10 – *Internet Protocols Part II and MPLS*: Domain Name System (DNS), Dynamic Host Configuration Protocol (DHCP), HTTP (Hypertext Transfer Protocol), Internet Control Message Protocol (ICMP), Multi-protocol Label Switching (MPLS), IP Multicast, Internet Group Management Protocol (IGMP), and Socket Programming

Chapter 11 – *Voice over Internet Protocol (Voice over IP)*: VoIP Operation, VoIP Protocol, Session Initiation Protocol (SIP), SIP Components, Connection Operation, and Bandwidth Calculation for VoIP

Chapter 12 – *Wireless Local Area Network (WLAN)*: WLAN Topologies, Wireless LAN Technology, WLAN Standards (IEEE 802.11 Families), Wireless LAN Physical Layers, IEEE 802.11g/n/ax, WLAN Medium Access, and MAC Frame Format

Chapter 13 – *Low Power Wireless Technologies for Internet of Things (IoT)*: ZigBee Operation and Components, ZigBee Topologies, ZigBee Application Profiles, ZigBee Protocol Architecture, Physical Layer, IEEE 802.15.4 MAC Layer ZigBee Security Modes, 6LoWPAN Architecture, LoRa Wide Area Network Technology (LoRa WAN), LoRaWAN Components, and LoRaWAN Security

Chapter 14 – *Introduction to Cryptography*: Elements of Network Security, Introduction to Cryptograph, RC4 Algorithm, Data Encryption Standard, Advanced Encryption Standard (AES) RSA Algorithm, Diffie-Hellman Key Exchange

Algorithm, Elliptic Curve Cryptography (ECC) Hash Value or Message Digest, Message Authentication Code (MAC), Digital Signatures, and Kerberos

Chapter 15 – *Network Security*: Secure Socket Layer Protocol (SSL/TLS), Virtual Private Network (VPN), IP Security Protocol (IPsec), Secure Shell (SSH), IEEE 802.1X, Extensible Authentication Protocol (EAP), Certificates, Firewalls, WLAN Security, IEEE 802.11i, WPA, WPA2, WPA3

New Haven, CT, USA   Ata Elahi
                                                                                   Alex Cushman

Acknowledgments

We would like to express our special thanks to Professor Lancor, Chairman of the Computer Science Department at Southern Connecticut State University for her support, and to Mr. Nicholas Brenckle for developing a program to find the points and the sum of points on an elliptic curve.

We wish to acknowledge and thank Ms. Mary E. James, Senior Editor in Applied Sciences, Professor Podnar, Mr. Omar Abid, and Vika Konovalenko for helping develop the manuscript for this text. Finally, we would like to thank the students of CSC 265 – Computer Networking and Security I and CSC 565 – Computer Networks for testing the initial versions of this textbook.

Contents

About the Authors

Ata Elahi is a Professor of Computer Science at Southern Connecticut State University. Dr. Elahi holds a Ph.D. in Electrical Engineering from Mississippi State University and is also the author of the following textbooks:

Elahi, A. *Computer Systems: Digital Design, Fundamentals of Computer Architecture and ARM Assembly Language*, 2nd edition, Springer, 2022

Elahi, A. *Computer Systems: Digital Design, Fundamentals of Computer Architecture and ARM Assembly Language*, Springer, 2018

Elahi, A., Arjeski, T. *ARM Assembly Language with Hardware Experiments*, Springer, 2015

Elahi, A., Gschwender, A. *ZigBee Wireless Sensor and Control Network*. Prentice Hall, 2010

Elahi, A., Elahi, M. *Data Network, & Internet Communications Technology*, Thomson Delmar Learning, 2006

Elahi, A. *Communication Network Technology*, Thomson Delmar Learning, 2001

Alex Cushman is a graduate student in Computer Science and Cybersecurity at Southern Connecticut State University with extensive professional experience in networking and security.

Chapter 1
Introduction to Communications Networks

Objectives
After completing this chapter, you should be able to:

- Explain the components of a Data Communication System.
- Explain the advantages of computer networks.
- Describe the components of a network.
- Discuss the function of a client/server model.
- Explain various networking topologies.
- Describe different types of networks in terms of their advantages and disadvantages.

1.1 Introduction

In order to transfer information from a source to a destination, some hardware components are required. Figure 1.1 shows the components of a data communication model for transmitting information from source to destination.

Source The source station can be a computer or server, and its function is to pass information to a transmitter for subsequent transmission.

Transmitter The function of a transmitter is to accept information from the source and change the information such that it is compatible with the transmission link. Information is then transmitted over the communication link. Modems and network cards are examples of transmitters.

Transmission Link The function of a transmission link is to carry information from a transmitter to a receiver. The transmission link can be a conductor, fiber optic cable, or wireless media (air).

© The Author(s), under exclusive license to Springer Nature Switzerland AG 2024
A. Elahi, A. Cushman, *Computer Networks*,
https://doi.org/10.1007/978-3-031-42018-4_1

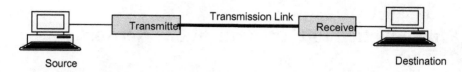

Fig. 1.1 A data communication model

Receiver The receiver accepts information from the transmission link. The information is then converted to proper form so that it is acceptable to the destination.

Destination The receiver passes information to the destination for processing.

In networking technology, both the receiver and transmitter come in one unit and are usually installed inside the computer, such as network cards or modems.

1.2 Computer Networks

Networking is a business tool for companies. For example, a bank can transfer funds between branches by using a network, people can access their bank accounts using automatic teller machines via a network, and travel agencies use networks to make airline reservations. Everyday online activities such as shopping and banking transactions are also possible because of computer networks. Students can now access the Internet in any location on their campus from their laptop computers or smartphone, thanks to the rapid growth of networking technology.

Networking is a generic term. Several computers connected to each other are called a **computer network**. A network is a system of computers and related equipment connected by communication links to share data. The related equipment may be printers, fax machines, modems, copiers, and so forth. The following are some of the benefits of using computer networks:

Resource Sharing: Computers in a network can share resources such as data, printers, disk drives, and scanners.

Reliability: Since computers in a network can share data, if one of the computers on the network crashes, a copy of its resources might be found on other computers in the network.

Cost: Microcomputers are much less expensive than mainframes. Instead of using several mainframes, a network can use one mainframe as a server, with several microcomputers connected to the server as clients. This creates a client/server relationship.

Communication: Users can exchange messages via electronic mail or other messaging systems, or they can transfer files.

1.3 Network Models

A computer in a network can be either a server, client, or a peer. A **server** is a computer on the network that holds shared files and the network operating system that manages the network operation. For a **client** computer to access the resources in the server, the client computer must request information from the server. The server will then transmit the information requested to the client.

The following three models are used, based on the type of network operation needed:

1. Peer-to-peer network (work group)
2. Server-based network
3. Client/server network

Peer-to-Peer Model (Work Group)

In a peer-to-peer model, there is no special station that holds shared files and a network operating system. Each station can access the resources of the other stations in the network. Individual stations can act as a server and/or a client. In this model, each user is responsible for administrating and upgrading the software of his or her station. Since there is no centralized station to manage network operation, this model is typically used for a network of fewer than ten stations. Figure 1.2 shows a peer-to-peer network model.

File Server Model

In a file server model, a server stores all the network's shared files and applications such as word processor documents, compilers, database applications, and the network operating system (NOS) that manages network operations. A user can access the file server and transfer shared files to his or her station.

Figure 1.3 shows a network with one file server and three users or clients. Each client can access the resources on the server and the resources of other clients. Clients that are connected on a network may be able to freely exchange information with one another.

Some of the most popular servers are as follows:

Mail server: A mail server stores all the clients' mail. The client can access the server and transfer incoming mail to its station. The client can also use the mail server to transfer outgoing mail to the mail server of another network.

Print server: Clients can submit files to the server for printing.

Fig. 1.2 Peer-to-peer network

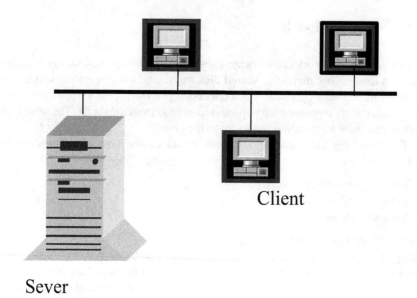

Fig. 1.3 Network with one server and three clients

Fig. 1.4 Client/server model

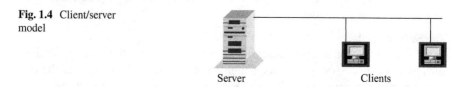

Communication server: The server is used by clients to communicate with other networks via communication links.

Client/Server Model

In the **client/server model**, a client first submits a task to the server. The server then executes the client's task and returns the results to the requesting client station. In comparison with the server-based model, the client/server model makes more efficient use of the network as it requires less information to travel through the network. This method of information sharing is depicted in Fig. 1.4.

1.4 Network Components

A network is composed of the following basic components:

1. **Network Interface Card (NIC):** Each computer in a network requires a Network Interface Card. The NIC allows the stations on the network to communicate with each other.

2. **Transmission Medium**: The transmission medium connects the computers together and provides a communication link between the computers on the network. Some of the more common types are twisted pair cable, coaxial cable, fiber optic cable, and wireless.
3. **Network Operating System (NOS)**: The NOS runs on the server and provides services to the client such as login, password, print file, network administration, and file sharing. Most modern computer operating systems have NOS functionality.

1.5 Network Topology

The **topology** of a network describes the way computers are connected. Topology is a major design consideration for cost and reliability. The following is a list of common topologies found in computer networking.

- Star
- Ring
- Bus
- Mesh
- Tree
- Hybrid

Star Topology
In a **star** topology, all stations are connected to a central controller or hub as shown in Fig. 1.5. For any station to communicate with another station, the source must send information to the hub, then the hub must transmit that information to the destination station. If station #1 wants to send information to station #3, it must send information to the hub and the hub must pass the information to station #3.

The disadvantage of the star topology is that the operation of the entire network depends on the hub. If the hub breaks down, the entire network is disabled. The advantages of star topology are as follows:

Fig. 1.5 Star topology

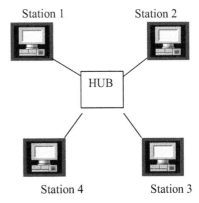

- It is easy to set up the network.
- It is easy to expand the network.
- If one link to the hub breaks, only the station using that link is affected.

It is possible for a network to have one topology electrically, or logically, but another topology physically. For example, Ethernet with unshielded twisted pair cabling uses the star topology physically, but the stations are connected logically using a bus topology.

Ring Topology

IBM invented the ring topology, which is well known as IBM Token Ring. In a **ring topology,** all stations are connected in cascading order to make a ring, as shown in Fig. 1.6. The source station transfers information to the next station on the ring, which checks the address of the information. If the address of the information matches with the station's address, the station copies the information and passes it to the next station. The next station repeats the process and passes the information on to the next station, and so on, until the information reaches the source station. The source then removes the information from the ring. The arrows in Fig. 1.6 indicate the direction in which the information flows.

The disadvantages of ring topology are as follows:

- If a link or a station breaks down, the entire network is disabled.
- Complex hardware is required (the network interface card is expensive).
- Adding a new client disrupts the entire network.

The advantages of ring topology are as follows:

- It is easy to install.
- It is easy to expand.

Fig. 1.6 Ring topology

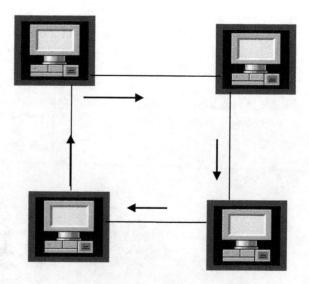

Bus Topology

A **bus network** is a multi-point connection in which stations are connected to a single cable called a bus. In the bus topology, all stations share one media as depicted in Fig. 1.7. The bus topology is one of the most popular topologies used in LAN networking and Ethernet is one of the most popular LANs that uses bus topology.

The advantages of bus topology are simplicity, low cost, and easy expansion of the network. The disadvantage of bus topology is that a breakdown in the bus cable brings the entire network down.

Mesh Topology

Mesh topology can be a **full mesh topology** (fully connected topology) or **partial mesh topology**. In a full mesh topology, each station is directly connected to every other station in the network, as shown in Fig. 1.8.

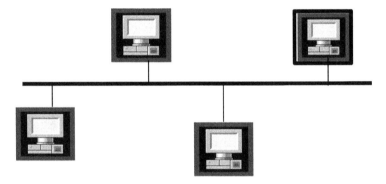

Fig. 1.7 Bus topology

Fig. 1.8 Full mesh topology

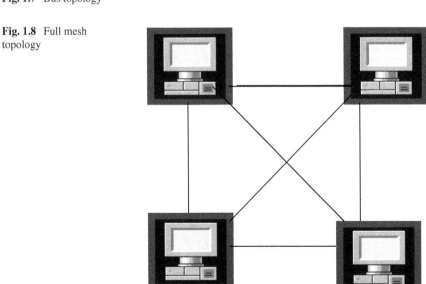

The advantage of a fully connected topology is that each station has a dedicated connection to every other station. Therefore, this topology offers the highest reliability and security. If one link in the mesh topology breaks, the network remains active.

A major disadvantage of a fully connected topology is that it uses many connections and therefore requires a great deal of wiring, especially when the number of stations increases. Consider, for example, a fully connected network with 100 workstations. Workstation #1 would require 99 network connections to connect it to workstations 2 through 100. The total number of connections is determined by N(N-1)/2, where N is the number of stations in the network. This type of topology is seldom used because it is not cost-effective.

In **partial mesh** topology, some stations are connected to many other stations, but others are connected only to those stations with which they exchange the most data. Figure 1.9 shows partial mesh topology.

Tree Topology

The tree topology uses an active hub or repeater to connect stations together. The **hub** is one of the most important elements of a network because it links stations in the network together. The function of the hub is to accept information from one station and repeat the information to other stations or hubs, as shown in Fig. 1.10.

The advantage of this topology is that when one hub breaks, only stations connected to the broken hub will be affected. There are several types of hubs as listed below.

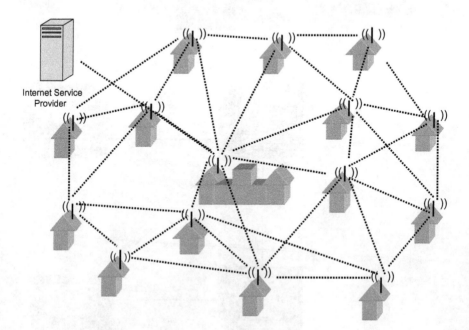

Fig. 1.9 Partial mesh topology

Fig. 1.10 Tree topology

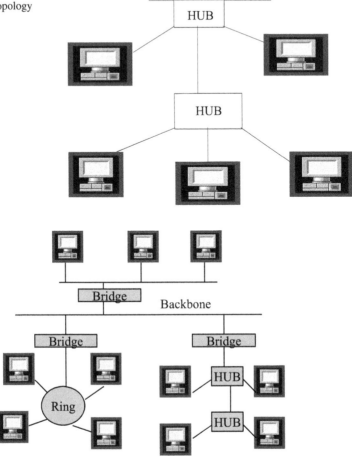

Fig. 1.11 Hybrid topology

Manageable Hub: Intelligent hubs are defined as manageable hubs, which means each of the ports on the hub can be enabled or disabled by the network administrator through software.

Stand-Alone Hub: A stand-alone hub is a type of hub used for workgroups of computers that are separate from the rest of the network. They cannot be linked together logically to represent a larger hub.

Modular Hub: A modular hub comes with a chassis or card cage and the number of ports can be extended by adding extra cards.

Stackable Hub: A stackable hub looks like a stand-alone hub, but several of them can be stacked or connected together in order to increase the number of ports.

Hybrid Topology

Hybrid topology is a combination of different topologies connected together by a backbone cable as shown in Fig. 1.11. Each network is connected to the backbone cable by a device called a bridge.

1.6   Types of Networks

The distance between computers that are connected as a network determines the type of network, such as a Local Area Network (LAN), Metropolitan Area Network (MAN), and Wide Area Network (WAN).

Local Area Network (LAN)
A **Local Area Network (LAN)** is a high-speed network designed to link computers and other data communication systems together within a small geographic area such as an office, department, or a single floor of a multi-story building. Several LANs can be connected together in a building or campus to extend the connectivity. A LAN is considered a private network. The most popular LANs in use today are Ethernet, Token Ring, and Gigabit Ethernet.

Metropolitan Area Network (MAN)
Metropolitan Area Networks (MAN) can cover approximately 30 to 100 miles, connecting multiple networks which are in different locations of a city or town. The communication links in a MAN are generally owned by a network service provider. Figure 1.12 shows a Metropolitan Area Network.

Wide Area Networks (WANs)
A **Wide Area Network (WAN)** is used for long-distance transmission of information. WANs cover a large geographical area, such as an entire country or continent. WANs may use leased lines from telephone companies, Public Switch Data Networks (PSDN), or satellites for communication links.

The **Internet** is a collection of globally scattered networks which are connected through gateways, as shown in Fig. 1.13. Each gateway has a routing table containing information about the networks to which it is connected as one or more networks may be connected to a single gateway. A gateway is designed to accept information from a source network and check its routing table to determine if the

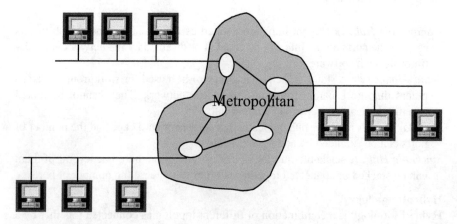

Fig. 1.12 Metropolitan Area Network

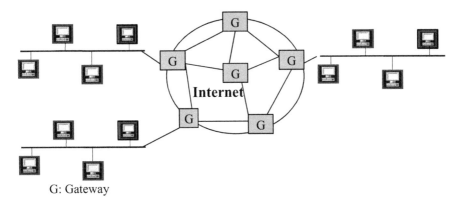

G: Gateway

Fig. 1.13 Internet architecture

destination station is in a network which is connected to the gateway. If the destination station is in a connected network, it transmits the information to the destination network. Otherwise, it passes the information to the next gateway, which performs the same operation. This process continues until the information reaches its destination.

Wireless Network Technologies
Wireless network technologies can be categorized based on power consumption. These include the following:

1. Wireless Local Area Networks (WLAN), used for offices and buildings
2. Low-Power Wireless Network technologies for Internet of Things (IoT)

   (a) Low-Power Wireless Personal Area Network (LoWPAN) Technologies with a coverage of up to 100 m such as ZigBee, 6LoPAN, Thread, and Z-Wave and Bluetooth
   (b) Low-Power Wide Area Networks (LPWAN) with coverage of kilometers, such as LoRa, Sigfox, and Dash

Internet Backbone and Internet Service Providers
The **backbone** of the Internet is composed of subterranean, transmission media which allow information to be transmitted globally. These transmission media are owned by national **Internet Service Providers (ISPs)** such as AT&T, Verizon, and Sprint. Figure 1.14 shows the part of the backbone owned by Verizon. To create the international internetwork known as the Internet, these ISPs are connected to the national ISPs of other countries. The transmission media owned by national ISPs are leased to Regional ISPs and Local ISPs which provide consumers with internet access in their homes and offices. Some Regional ISPs in the United States include Frontier Communications and XFINITY Comcast as they only provide certain states with internet access. Local ISPs may be any office or university which provides its users with internet access.

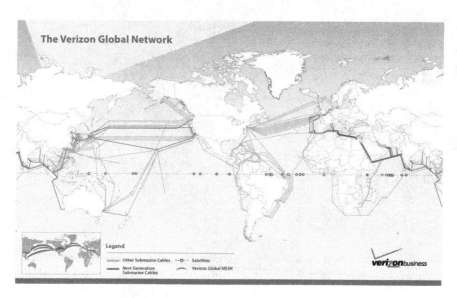

Fig. 1.14 Verizon Internet backbone

1.7   Communication Protocols and Standard Organizations

A **communication protocol** is a set of rules used by computers which allows them
to communicate with each other. Computers must follow certain rules in order to
communicate with each other. Some of the rules that define a protocol are as follows:

- *Size of information.* Both computers must agree on the minimum and maximum
  size of information.
- *How to represent information.* Information may be Unicode, ASCII, or encrypted.
- *Error detection.* The method used by the receiver to check the integrity of
  information.
- *Receipt of information.* The transmitter must know that information has been
  received at the destination.
- *Non-receipt of information.* Both computers must know what to do if information sent is not received or if it is received but has been corrupted.

Some of the common network protocols are:

TCP/IP: Transmission Control Protocol/Internet Protocol used in the Internet and
many LANs.
NetBEUI: NetBIOS Extended User Interface is a small and fast protocol used for
small LANs.
X.25: X.25 is a set of protocols used in packet switching networks.
IPX/SPX: Novell NetWare uses Internet Packet Exchange/Sequenced Packet
Exchange.

NWlink: NWlink is a Microsoft version of IPX/SPX.
AppleTalk: AppleTalk developed by Apple for MAC.

1.7.1   Standard Organizations

There are several organizations that are constantly working toward developing standards for computers and communication equipment. The development of standards for computers enables hardware and software products made by different vendors to be compatible. Standardization allows products from different manufacturers to work together in creating customized systems. Without standards, only hardware and software made by the same manufacturer can work properly together. The following is a list of **standards organizations**:

IEEE   The Institute for Electrical and Electronics Engineers (IEEE) is the largest technical organization in the world. The objective of IEEE is to advance the field of electronics, computer science, and computer engineering. The IEEE also develops standards for computers, electronics, and local area networks (in particular, the IEEE 802 standards).

ITU   The International Telecommunication Union (ITU) was founded in 1864 and became a United Nations Agency with the purpose of defining standards for telecommunications, Wide Area Networks (WAN), Asynchronous Transfer Mode (ATM), and Integrated Services Digital Networks (ISDN).

EIA   The Electrical Industry Association (EIA) is a trade association representing high technology manufacturers in the United States. The EIA develops standards for connectors and transmission media. Some of the well-known EIA standards are RS-232 and RJ-45.

ANSI   The American National Standards Institute was founded in 1918. ANSI is composed of 1300 members representing computer companies, with the purpose of developing standards for the computer industry. ANSI is the US representative in the International Organization for Standardization (ISO). Some of the well-known ANSI standards include optical cabling, programming languages (ANSI C), and the Fiber Distributed Data Interface (FDDI).

ISO   The International Organization for Standardization (ISO) is an international organization that comprises national standards bodies of seventy-five countries. The ISO develops standards for a wide range of products, including the model for networks called the Open System Interconnection (OSI) model.

IETF   The Internet Engineering Task Force (IETF) develops standards for the Internet, such as Internet Protocol version 6 (IPv6), HTTP, and DNS. The IETF is composed of international network designers, network industries, and researchers.

World Wide Web Consortium (*W3C*)   The World Wide Web Consortium develops standards for web technologies, such as HTML.

1.8   Networking Protocol Models

Two common protocol models, or suites, in use today are the 5-layer **TCP/IP (Transmission Control Protocol/Internet Protocol)** model and the 7-layer **OSI (Open Systems Interconnections)** model.

TCP/IP Protocol Suite   The TCP/IP Protocol suite consists of the following five layers as shown in Fig. 1.15.

Layer 1: Physical Layer
The **Physical layer** defines the type of signal and type of connectors (such as RS-232 or RJ-45) to be used for the Network Interface Card (NIC). It defines cable types (such as coaxial cable, twisted pair or fiber-optic cable) to be used for the transmission media. It accepts incoming signals from the media and converts those signals bits and converts outgoing data bits to signals for the transmission over media.

Layer 2: Data Link Layer
The **Data Link layer** defines the frame format, such as the start of a frame, end of a frame, size of a frame, and type of transmission. The Data Link layer performs the following functions:

1. *On the transmitting side*: The Data Link layer accepts information from the Network layer and breaks the information into frames. It then adds the destination MAC address, source MAC address, and Frame Check Sequence (FCS) field, then passes each frame to the Physical layer for transmission.
2. *On the receiving side*: The Data Link layer accepts the bits from the Physical layer and forms them into a frame, performing error detection. If the frame is free of errors, the Data Link layer passes the frame up to the Network layer.
3. *Frame synchronization*: This layer identifies the beginning and end of each frame.
4. *Flow control:* Distinguishes between control frames and information frames.
5. *Link management*: It coordinates transmission between the transmitter and the receiver.
6. *Determine contention method*: It defines an access method in which two or more network devices compete for permission to transmit information across the same communication media, such as in token passing and Carrier Sense Multiple Access with Collision Detection (CSMA/CD).

Layer 3: Network Layer
The function of the **Network layer** is to perform routing. Routing determines the route, or pathway, for moving information in a network with multiple LANs. The Network layer checks the logical address of each frame and forwards the frame to

Fig. 1.15   TCP/IP model

Layer 5	Application
Layer 4	Transport
Layer 3	Network
Layer 2	Data link
Layer 1	Physical

the next router based on a routing table. The Network layer is responsible for translating each logical address (name address) to a physical address (MAC address). An example of a Network layer protocol is Internet Protocol (IP).

The Network layer provides two types of services: connectionless and connection- oriented services. In connection-oriented services, the Network layer makes a connection between a source and a destination, then starts transmission. In a connectionless service, there is no connection between source and destination. The source transmits information regardless of whether the destination is ready or not. A common example of this is e-mail.

Layer 4: Transport Layer

The **Transport layer** provides the reliable transmission of data in order to ensure that each frame reaches its destination. If, after a certain period of time, the Transport layer does not receive an acknowledgment from the destination, it retransmits the frame and again waits for acknowledgment from the destination. An example of a Transport layer protocol is Transmission Control Protocol (TCP).

Layer 5: Application Layer

The Application layer enables users to access the network with applications such as E-mail, FTP (File Transfer Protocol), HTTP (Hyper Text Transfer Protocol), and Telnet.

OSI Model   The OSI model is like the TCP/IP model, but with two additional layers between the Application and Transport layers, as shown in Fig. 1.16.

Layer 5: Session Layer

The **Session layer** establishes a logical connection between the applications of two computers that are communicating with each other. It allows two applications on two different computers to establish and terminate a session. When a workstation connects to a server, the server performs the login process, requesting a username and password. This is an example of establishing a session.

Layer 6: Presentation Layer

The **Presentation layer** receives information from the Application layer and converts it to a form acceptable by the destination. The Presentation layer converts information to ASCII, or Unicode, or encrypts or decrypts the information.

Layer 7: Application Layer

The Application layer enables users to access the network with applications such as E-mail, FTP (File Transfer Protocol), HTTP (Hyper Text Transfer Protocol), and Telnet.

Fig. 1.16 OSI model

Layer 7	Application
Layer 6	Presentation
Layer 5	Session
Layer 4	Transport
Layer 3	Network
Layer 2	Data link
Layer 1	Physical

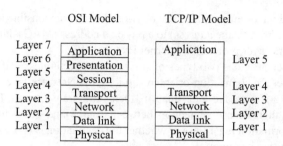

Fig. 1.17 OSI and TCP/IP model comparison

Comparing Models Essentially, the first four layers of the OSI and TCP/IP models serve the same purpose. However, the Application layer of the TCP/IP model performs all the duties of top three layers of the OSI model as seen in Fig. 1.17.

Summary

- A group of several computers connected by communication media is termed a computer network.
- Some of the applications of computer networks are file sharing, resource sharing, media streaming, and e-mail.
- In the client/server model of networking, the client submits the information to the server, which processes the information and returns the results to the client station.
- The components of a network are a Network Interface Card (NIC), Network Operating System (NOS), and the communication link (transmission medium).
- Computers can be connected in the form of Star, Ring, Bus, Mesh, Tree, and Hybrid Topologies.
- The types of networks are LAN, MAN, WAN, and the Internet.
- Some of the standard organizations that develop standards for networks and data communications are the IEEE, ITU, EIA, ISO, IETF, and ANSI.
- A communication protocol is a set of rules used by two computers in order to communicate which each other.
- The most popular communication protocols are Transmission Control Protocol/Internet Protocol (TCP/IP), NetBEUI, IPX/SPX NWlink, and AppleTalk.
- The International Standard Organization (ISO) developed a model for networks called the Open System Interconnection (OSI) model.
- The OSI model consists of seven layers, from top to bottom: Application Layer, Presentation layer, Session layer, Transport layer, Network layer, Data Link layer and Physical layer.
- The Application layer enables the user to access the network applications.
- The Presentation layer is responsible for the representation of information such as ASCII encoded data, encryption, and decryption.

- The function of the Session layer is to establish a session between a source application and destination application and to disconnect a session between two applications.
- The function of the Transport layer is to ensure that data gets to the destination, to perform error control and flow control, and to assure quality of service.
- The function of the Network layer is to deliver information from a source to its destination by routing that information.
- The function of the Data Link layer is framing, error detection, and retransmission.
- The functions of the physical layer are establishing the electrical interface for transmission and the type of the signal, and the conversion of logical bits to electronic signals and vice versa.

Key Terms

Bus Network	OSI Model
Client	Peer-to-Peer Model
Client/Server Model	Ring Topology
Computer Network	Server
File Server Model	Standard Organization
Hybrid Topology	Star Topology
Internet	TCP/IP Model
Local Area Network (LAN)	Topology
Mesh	Transmission Medium
Metropolitan Area Network (MAN)	Tree Topology
Network Interface Card (NIC)	Wide Area Network (WAN)
Network Operation System (NOS)	

Review Questions

Multiple Choice Questions

1. Several computers connected together are called a _____.

    (a) Computer network
    (b) Client
    (c) Server
    (d) Hub

2. In a _____ network, the client submits a task to the server, then the server executes and returns the result to the requesting client station.

    (a) Peer-to-peer
    (b) Server based
    (c) Client/server
    (d) All of the above

3. A computer in a network can function as a _____ or as a _____.

   (a) Client, server
   (b) Client, user
   (c) a and b
   (d) All of the above

4. A _____ stores all the client's mail.

   (a) File server
   (b) Print server
   (c) Communication server
   (d) Mail server

5. A _____ uses a modem or other type of communication link to enable clients to communicate with other networks.

   (a) Mail server
   (b) Communication server
   (c) a and b
   (d) None of the above

6. In a _____ topology, all stations are connected to a central controller or hub.

   (a) Star
   (b) Ring
   (c) Bus
   (d) Mesh

7. In a _____ topology, all stations are connected in cascade.

   (a) Star
   (b) Ring
   (c) Tree
   (d) Bus

8. A _____ topology is a combination of different topologies connected together by a backbone cable.

   (a) Star
   (b) Ring
   (c) Bus
   (d) Hybrid

9. Which network topology uses a hub? _____

   (a) Ring
   (b) Bus
   (c) Star
   (d) Mesh

10. Which type of topology uses multi-point connections? _____

(a) Bus
(b) Star
(c) Ring
(d) Full mesh

11. How many connections are required by a fully connected network with five stations? _____.

(a) 5
(b) 10
(c) 20
(d) 15

12. Which of the following networks is used for office buildings? _____.

(a) LAN
(b) MAN
(c) WAN
(d) Internet

13. Which of the following topologies is used for Ethernet? _____.

(a) Bus
(b) Star
(c) Ring
(d) a and b

14. The Internet is a collection of LANs connected together by _____.

(a) Routers
(b) Switches
(c) Gateways
(d) Repeaters

15. Computers on a campus are connected by a/an _____.

(a) LAN
(b) WAN
(c) MAN
(d) Internet

16. The IEEE developed the _____ standard for LAN.

(a) IEEE 802
(b) RS232
(c) OSI Model
(d) All of the above

17. The _____ defines standards for telecommunications.

    (a) IEEE
    (b) ITU
    (c) EIA
    (d) ISO

18. The _____ defines standards for programming languages.

    (a) IEEE
    (b) ISO
    (c) ANSI
    (d) IETF

19. The _____ protocol is used on the Internet.

    (a) TCP/IP
    (b) X.25
    (c) IPX/SPX
    (d) NWLink

20. Microsoft's version of IPX/SPX is called _____.

    (a) Net BEUI
    (b) TCP/IP
    (c) NWLink
    (d) X.25

21. The OSI model contains _____ layers.

    (a) 4
    (b) 3
    (c) 7
    (d) 6

22. The _____ layer establishes a connection.

    (a) Network
    (b) Physical
    (c) Data Link
    (d) Application

23. The _____ layer defines the format of the frame.

    (a) Transport layer
    (b) Data Link layer
    (c) Network layer
    (d) Physical layer

24. Which layer of the OSI model is responsible for forming a frame?

    (a) Data link
    (b) Transport
    (c) Session
    (d) Physical

25. Which layer of the OSI model performs encryption? _____

    (a) Session layer
    (b) Presentation layer
    (c) Data Link layer
    (d) Transport layer

26. The function of the network layer is_____.

    (a) Error detection
    (b) Routing
    (c) To set up a session
    (d) Encryption

27. Which layer of the OSI model converts electrical signals to bits?

    (a) Physical
    (b) Data link
    (c) Network
    (d) Application

28. Which layer determines the route for packets transmitted from source to destination? _____.

    (a) Data Link
    (b) Network
    (c) Transport
    (d) Physical

Short Answer Questions

1. What are the components of a communication model?
2. Explain the function of a server.
3. What is the function of the client in a file server model?
4. Explain the term "client/server model."
5. What are the advantages of a client/server model?
6. What are the three components of a network?
7. A Network Operating System runs on the _____.
8. List the six networking topologies.
9. What is the disadvantage of a fully connected mesh topology?
10. What is a hub?
11. What are the three types of area networks?
12. What does MAN stand for?

13. What is the Internet?
14. What does WAN stand for?
15. What are the advantages of the bus topology?
16. List the layers of the OSI Model.
17. List the layers of the TCP/IP Model.
18. List three applications of TCP/IP Model.
19. What is the function of the Transport layer?
20. What layer deals with frames?
21. What layer converts bits to electronic signals?

Chapter 2
Data Communications

Objectives

After completing this chapter, you should be able to:

- Distinguish between analog and digital signals.
- Distinguish between periodic and non-periodic signals.
- Convert decimal numbers to binary and hexadecimal and vice versa.
- Represent characters and decimal numbers in the 7-bit ASCII code.
- Compare serial, parallel, asynchronous, and synchronous transmission.
- List the communication modes.
- Explain the different types of digital encoding methods.
- Calculate a Block Check Character (BCC).
- Calculate a Frame Check Sequence (FCS).
- Learn different error detection methods.

2.1   Introduction

In order to understand network technology, it is important to know how information is represented for transmission from one computer to another. Information can be transferred between computers in one of two ways: an analog signal or a digital signal.

A. Elahi, A. Cushman, *Computer Networks*,
https://doi.org/10.1007/978-3-031-42018-4_2

2.2   Analog Signals

An analog signal is a signal whose amplitude is a function of time and changes gradually as time changes. Analog signals can be classified as non-periodic and periodic signals.

Non-periodic Signal   In a non-periodic signal, there is no repeated pattern in the signal as shown in Fig. 2.1.

Periodic Signal   A signal that repeats a pattern within a measurable time period is called a periodic signal, and the completion of a full pattern is called a *cycle*. The simplest periodic signal is a sine wave, which is shown in Fig. 2.2. In the time domain, a sine wave's amplitude $a(t)$ can be represented mathematically as $a(t) = A \sin(\omega t + \theta)$, where A is the maximum amplitude, ω is the angular frequency, and θ is the phase angle.

A periodic signal can also be represented in the frequency domain, where the horizontal axis is the frequency and the vertical axis is the amplitude of signal. Figure 2.3 shows the frequency domain representation of a sine wave signal.

Usually an electrical signal representing voice, temperature or a musical sound, is made of multiple waveforms. These signals have one fundamental frequency and multiple frequencies that are called harmonics.

Characteristics of an Analog Signal

The characteristics of a periodic analog signal are frequency, amplitude, and phase.

Frequency   Frequency (F) is the number of cycles in one second, $F = \dfrac{1}{T}$, and is represented in Hz (Hertz). If each cycle of an analog signal is repeated every one second, the frequency of the signal is one Hz. If each cycle of an analog signal is repeated 1000 times every second (once every millisecond), the frequency is:

$$f = \frac{1}{T} = \frac{1}{10^{-3}} = 1000 Hz = 1kHz$$

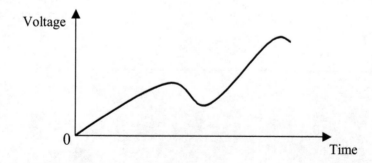

Fig. 2.1   Representation of a non-periodic analog signal

Fig. 2.2 Time domain representation of a sine wave

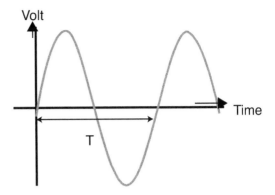

Fig. 2.3 Frequency representation of a sine wave

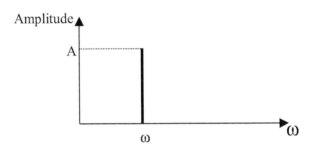

Table 2.1 shows different values for frequency and their corresponding periods.

Amplitude The amplitude of an analog signal is a function of time (as shown in Fig. 2.4) and may be represented in volts (unit of voltage). In other words, the amplitude is a signal's voltage value at any given time. At the time of t_1, the amplitude of signal is V_1.

Phase Two signals with the same frequency can differ in phase. This means that one of the signals starts at a different time from the other one. This difference can be measured in degrees, from 0 to 360 degrees, or in radians, where $360° = 2\pi$ radians. A sine wave signal can be represented by the equation $a(t) = A\,Sin(\omega t + \theta)$, where A is the peak amplitude, ω (omega) is the frequency in radians per second, t is the time in seconds, and θ is the phase angle. Cyclic frequency f can be expressed in terms of ω according to $f = \dfrac{\omega}{2\pi}$. A phase angle of zero means that the sine wave starts at time $t = 0$ and a phase angle of 90 degrees means that the signal starts at 90 degrees as shown in Fig. 2.5.

Example 2.1 Find the equation for a sine wave signal with frequency of 10 Hz, maximum amplitude of 20 volts, and phase angle of zero.

Table 2.1 Typical units of frequency and period

Units of frequency	Numerical value	Units of period	Numerical value
Hertz (Hz)	$1\,Hz$	Second (s)	1 s
Kilo Hertz (kHz)	$10^3\,Hz$	Millisecond (ms)	10^{-3} s
Mega Hertz (MHz)	$10^6\,Hz$	Microsecond (µs)	10^{-6} s
Giga Hertz (GHz)	$10^9\,Hz$	Nanosecond (ns)	10^{-9} s
Tera Hertz (THz)	$10^{12}\,Hz$	Picosecond (ps)	10^{-12} s

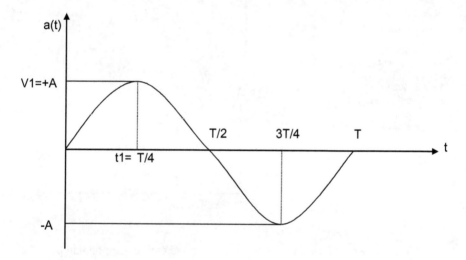

Fig. 2.4 A sine wave signal over one cycle

$$\omega = 2\pi f = 2\times 3.1416\times 10 = 62.83\,\frac{\text{rad}}{\text{sec}}$$
$$a(t) = 20\sin(62.83t)$$

2.3   Digital Signals

Modern computers communicate by using digital signals. **Digital signals** are represented by two voltages: one voltage represents the number 0 in binary, and the other
voltage represents the number 1 in binary. An example of a digital signal is shown
in Fig. 2.6, where 0 volts represents 0 in binary and + 5 volts represents 1.

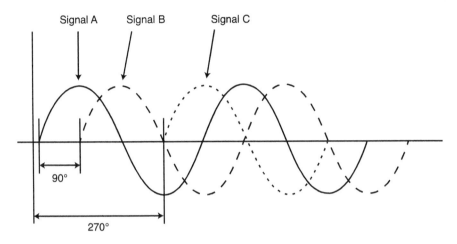

Fig. 2.5 Three sine waves with different phases

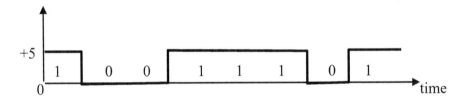

Fig. 2.6 Digital signal

2.4   Binary Numbers

Binary, or Base-2 numbers, are represented by 0 and 1. A binary digit, 0 or 1, is called a **bit**. Eight bits are equal to one **byte**. Two or more than two bytes are called a **word**. The hexadecimal number system has a base of 16 and, therefore, has 16 symbols (0 through 9 and A through F). Table 2.2 shows the decimal numbers, their binary values from 0 to 15, and their hexadecimal equivalents.

Converting from Hex to Binary
Table 2.2 can also be used to convert a number from hexadecimal to binary and from binary to hexadecimal.

Example 2.2 Convert the binary number 001010011010 to hexadecimal. Each 4 bits are grouped from right to left. By using Table 2.2, each 4 bit group can be converted to its hexadecimal equivalent.

0010	1001	1010
2	**9**	**A**

Example 2.3 Convert $(3D5)_{16}$ to binary. By using Table 2.2, the result in binary is

3	D	5
0011	**1101**	**0101**

The resulting binary number is 001111010101.

Example 2.4 Convert 6DB from hexadecimal to binary. By using Table 2.2, the result in binary is

6	D	B
0110	**1101**	**1011**

The resulting binary number is 011011011011.

Converting from Binary to Decimal

In general, any binary number can be represented by Eq. 2.1.

$$(a_5\,a_4\,a_3\,a_2\,a_1\,a_0.a_{\,1}\,a_{\,2}\,a_{\,3})_2 \qquad\qquad (2.1)$$

where a_i is a binary digit or bit (either 0 or 1).
Equation 2.1 can be converted to decimal number by using Eq. 2.2.

Table 2.2 Decimal numbers with binary and hexadecimal equivalents

Decimal	Binary (base 2)	Hexadecimal (Base 16) or HEX
0	0000	0
1	0001	1
2	0010	2
3	0011	3
4	0100	4
5	0101	5
6	0110	6
7	0111	7
8	1000	8
9	1001	9
10	1010	A
11	1011	B
12	1100	C
13	1101	D
14	1110	E
15	1111	F

$$\underbrace{a_5\,a_4\,a_3\,a_2\,a_1\,a_0}_{\text{Interger}}\ \underbrace{a_1\,a\quad a}_{\text{Fraction}}{}_2 \qquad a_0\ 2^0\quad a_1\ 2^1\quad a_2\ 2^2\quad a_3\ 2^3$$

$$a_1\ 2^1\quad a\quad 2\quad \cdot{}_{..}$$

Example 2.5 To convert $(110111.101)_2$ to decimal:

$$110111.101_2\quad 1\,2^0\ \ 1\,2^1\ \ 1\,2^2\ \ 0\,2^3\ \ 1\,2^4\ \ 1\,2^5\ \ 1\,2^1\ \ 0\,2\ \ \ 1\,2\qquad 55.625$$

2.5 Coding Schemes

Since computers can only understand binary numbers (0 or 1), all information (such as numbers, letters, and symbols) must be represented as binary data. One commonly used code to represent printable and non-printable characters is the American Standard Code for Information Interchange (ASCII).

ASCII Code

Each character in ASCII code is represented by 8 bits where the most significant bit is used for a parity bit. Table 2.3 shows the **ASCII code** and its hexadecimal equivalent. Characters from hexadecimal 00 to 1F and 7F are control characters which are nonprintable characters, such as NUL, SOH, STX, ETX, ESC, and DLE (data link escape).

Example 2.6 Convert the word "Network" to binary and show the result in hexadecimal. By using Table 2.3, each character is represented by seven bits and results in

1001110	1100101	1110100	1110111	1101111	1110010	1101011
N	e	t	w	o	r	k

or in hexadecimal

4E   65   74   77   6F   72   6B

Universal Code or Unicode

Unicode is a new 16-bit character encoding standard for representing characters and numbers in most languages such as Greek, Arabic, Chinese, and Japanese. The

Table 2.3 American Standard Code for Information Interchange (ASCII)

Binary	Hex	Char	Binary	Hex	Char	Binary	Hex	Char	Binary	Hex	Char
0000000	00	NUL	0100000	20	SP	1000000	40	@	1100000	60	'
0000001	01	SOH	0100001	21	!	1000001	41	A	1100001	61	a
0000010	02	STX	0100010	22	"	1000010	42	B	1100010	62	b
0000011	03	ETX	0100011	23	#	1000011	43	C	1100011	63	c
0000100	04	EOT	0100100	24	$	1000100	44	D	1100100	64	d
0000101	05	ENQ	0100101	25	%	1000101	45	E	1100101	65	e
0000110	06	ACK	0100110	26	&	1000110	46	F	1100110	66	f
0000111	07	BEL	0100111	27	'	1000111	47	G	1100111	67	g
0001000	08	BS	0101000	28	(	1001000	8	H	1101000	68	h
0001001	09	HT	0101001	29	)	1001001	49	I	1101001	69	i
0001010	0A	LF	0101010	2A	*	1001010	4A	J	1101010	6A	j
0001011	0B	VT	0101011	2B	+	1001011	4B	K	1101011	6B	k
0001100	0C	FF	0101100	2C	,	1001100	4C	L	1101100	6C	l
0001101	0D	CR	0101101	2D	-	1001101	4D	M	1101101	6D	m
0001110	0E	SO	0101110	2E	.	1001110	4E	N	1101110	6E	n
0001111	0F	SI	0101111	2F	/	1001111	4F	O	1101111	6F	o
0010000	10	DLE	0110000	30	0	1010000	50	P	1110000	70	p
0010001	11	DC1	0110001	31	1	1010001	51	Q	1110001	71	q
0010010	12	DC2	0110010	32	2	1010010	52	R	1110010	72	r
0010011	13	DC3	0110011	33	3	1010011	53	S	1110011	73	s
0010100	14	DC4	0110100	34	4	1010100	54	T	1110100	74	t
0010101	15	NACK	0110101	35	5	1010101	55	U	1110101	75	u
0010110	16	SYN	0110110	36	6	1010110	56	V	1110110	76	v
0010111	17	ETB	0110111	37	7	1010111	57	W	1110111	77	w
0011000	18	CAN	0111000	38	8	1011000	58	X	1111000	78	x
0011001	19	EM	0111001	39	9	1011001	59	Y	1111001	79	y
0011010	1A	SUB	0111010	3A	:	1011010	5A	Z	1111010	7A	z
0011011	1B	ESC	0111011	3B	;	1011011	5B	[	1111011	7B	[
0011100	1C	FS	0111100	3C	<	1011100	5C	\	1111100	7C	/
0011101	1D	GS	0111101	3D	=	1011101	5D	]	1111101	7D	}
0011110	1E	RS	0111110	3E	<	1011110	5E	^	1111110	7E	~
0011111	1F	US	0111111	3F	?	1011111	5F	-	1111111	7F	DEL

ASCII code uses eight bits to represent each character in Latin, and it can represent 256 characters. The ASCII code does not support mathematical symbols and scientific symbols. **Unicode** uses 16 bits, which can represent 65,536 characters or symbols. A character in Unicode is represented by a 16-bit binary, equivalent to four digits in hexadecimal. For example, the character B in Unicode is U0042H (U represents Unicode). The ASCII code is represented between $(00)_{16}$ and $(FF)_{16}$. For converting ASCII code to Unicode, two zeros are added to the left side of ASCII code; therefore, the Unicode to represent ASCII characters is between $(0000)_{16}$ and $(00FF)_{16}$. Table 2.4 shows some of the Unicode for Latin and Greek characters. Unicode is divided into blocks of code, with each block assigned to a specific language. Table 2.5 shows each block of Unicode for some different languages.

Table 2.4   Unicode values for some Latin and Greek characters

Latin		Greek	
Character	Code (Hex)	Character	Code (Hex)
A	U0041	φ	U03C6
B	U0042	α	U03B1
C	U0043	γ	U03B3
0	U0030	μ	U03BC
8	U0038	β	U03B2

Table 2.5   Unicode block allocations

Start code (Hex)	End code (Hex)	Block name
U0000	U007F	Basic Latin
U0080	U00FF	Latin supplement
U0370	U03FF	Greek
U0530	U058F	Armenian
U0590	U05FF	Hebrew
U0600	U06FF	Arabic
U01A0	U10FF	Georgian

2.6   Transmission Modes

When data is transferred from one computer to another by digital signals, the receiving computer has to distinguish the size of each signal to determine when a signal ends and when the next one begins. For example, when a computer sends a signal as shown in Fig. 2.7, the receiving computer has to recognize how many ones and zeros are in the signal. Synchronization methods between source and destination devices are generally grouped into two categories: asynchronous and synchronous.

Asynchronous Transmission
Asynchronous transmission occurs character by character and is used for serial communication, such as by a modem or serial printer. In asynchronous transmission, each data character has a start bit which identifies the start of the character, and one or two bits which identifies the end of the character, as shown in Fig. 2.8. The data character is 7 bits. Following the data bits may be a parity bit, which is used by the receiver for error detection. After the parity bit is sent, the signal must return to high for at least one bit to identify the end of the character. The new start bit serves as an indicator to the receiving device that a data character is coming and allows the receiving side to synchronize its clock. Since the receiver and transmitter clocks are not synchronized continuously, the transmitter uses the start bit to reset the receiver clock so that it matches the transmitter clock. Also, the receiver is already programmed for the number of bits in each character sent by the transmitter.

Synchronous Transmission
Some applications require transferring large blocks of data, such as a file from disk or transferring information from a computer to a printer. **Synchronous**

Fig. 2.7 Digital signal

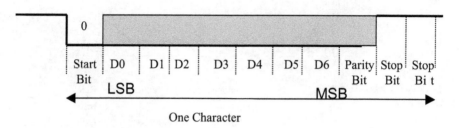

Fig. 2.8 Asynchronous transmission

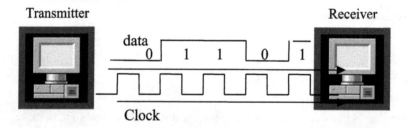

Fig. 2.9 Synchronous transmission

transmission is an efficient method for transferring large blocks of data by using time intervals for synchronization.

One method of synchronizing a transmitter and receiver is through the use of an external connection which carries a clock pulse. The clock pulse represents the data rate of the signal, as shown in Fig. 2.9, and is used to determine the speed of data transmission. The receiver of Fig. 2.9 reads the data as 01101.

Figure 2.9 shows that an extra connection is required to carry the clock pulse for synchronous transmission. In networking, one medium is used for transmission of both information and the clock pulse. The two signals are encoded in a such way that the synchronization signal is embedded into the data. This can be done with Manchester encoding or Differential Manchester encoding.

2.7 Transmission Methods

There are two types of transmission methods used for sending digital signals from one station to another across a communication channel: serial transmission and parallel transmission.

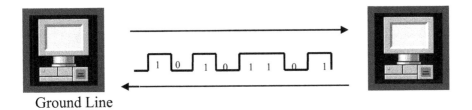

Ground Line

Fig. 2.10 Serial transmission

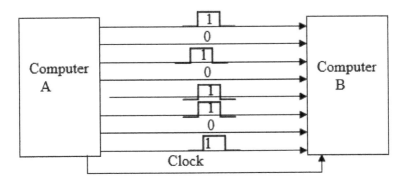

Fig. 2.11 Parallel transmission

Serial Transmission
In **serial transmission**, information is transmitted one bit at a time over one wire as shown in Fig. 2.10.

Parallel Transmission
In **parallel transmission**, multiple bits are sent simultaneously, one byte or more at a time, instead of bit by bit as in serial transmission. Figure 2.11 shows how computer A sends eight bits of information to computer B at the same time by using eight different wires. Parallel transmission is faster than serial transmission, at the same clock speed.

2.8   Communication Modes

A communication mode specifies the capability of a device to send and receive data by determining the direction of the signal between two connections. There are three types of communication modes: simplex, half-duplex, and full-duplex.

Simplex Mode
In **simplex mode**, the transmission of data goes in one direction only, as shown in Fig. 2.12. A common analogy is a commercial radio or TV broadcast where the sending device never requires a response from the receiving device.

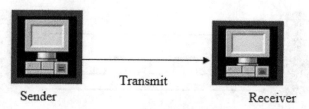

Fig. 2.12 Simplex transmission

Fig. 2.13 Half-duplex
transmission

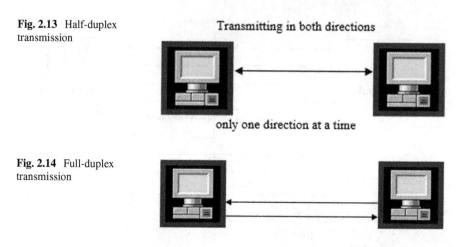

Fig. 2.14 Full-duplex
transmission

Half-Duplex Mode

In **half-duplex** mode, two devices exchange information as shown in Fig. 2.13; however, information can be transmitted across the channel one direction at a time. A common example is Citizen Band radio (CB) or ham radio where a user can either talk or listen, but both parties cannot talk at the same time.

Full-Duplex Mode

In **full-duplex** mode, both computers can send and receive information simultaneously, as shown in Fig. 2.14. An example of full-duplex is our modern telephone system, in which both users may talk and listen at the same, with their voices carried two ways simultaneously over the phone lines.

2.9   Signal Transmission

There are two methods used to transfer information over media: baseband and broadband transmission.

Baseband Transmission Mode

When the entire bandwidth of a cable is used to carry only one signal, the cable operates in **baseband** mode. Many digital signals use baseband transmission.

Broadband Transmission Mode
When the bandwidth of a cable is used to carry several signals simultaneously, the cable operates in **broadband mode**. For example, cable TV transmission works in broadband mode because it carries multiple channels using multiple signals over the cable. Broadband mode frequently uses analog signals.

2.10   Digital Signal Encoding

Digital signal encoding is used to represent binary values in the form of digital signals. The receiver of the digital signal must know the timing of each signal, such as the start and end of each bit. Following are some methods used to represent digital signals:

- Unipolar encoding
- Polar encoding
- Bipolar encoding
- Non-return to zero (NRZ)
- Non-return to zero inverted (NRZ-I)
- Manchester and differential Manchester encoding

Manchester and differential Manchester, and non-return to zero inverted (NRZ-I) encoding schemes are used in LANs and non-return to zero is used in WANs. Each encoding technique is described below.

Unipolar Encoding
In **unipolar encoding,** only positive voltage or negative voltage are used to represent binary 0 and 1. For example, + 5 volts represents binary 1 and zero volts represents 0, as shown in Fig. 2.15.

Polar Encoding
In **polar encoding,** positive and negative voltages are used to represent binary one and zero, respectively. For example, +5 volts represents binary one and −5 volts represent binary zero, as shown in Fig. 2.16.

Bipolar Encoding
In **bipolar encoding**, signal voltage varies in three levels: positive, zero, and negative voltage. One of the most popular bipolar encoding methods is Alternate Mark

Fig. 2.15 Unipolar encoding signals

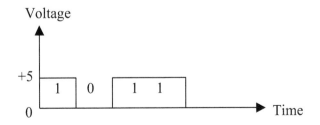

Fig. 2.16 Polar encoding
signals

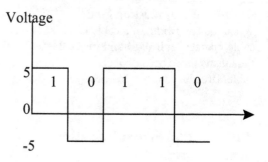

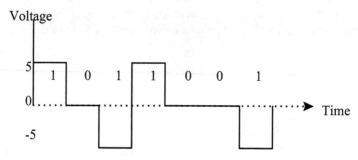

Fig. 2.17 Bipolar encoding signals

Inversion (AMI). In AMI encoding, binary 0 is represented by zero volts and binary 1 is represented by alternating swings between positive and negative voltages, as shown in Fig. 2.17.

Non-return to Zero Encoding (NRZ)
NRZ is a simple format of polar encoding, which uses two voltage levels for representing 1 and 0, with binary 0 represented by a positive voltage and binary 1 represented by a negative voltage, as shown in Fig. 2.18.

Non-return to Zero Inverted Encoding (NRZ-I)
In NRZ-I, there is a transition at the start of logic 1 (low to high or high to low) and no transition at start of 0, as shown in Fig. 2.19.

Manchester and Differential Manchester Encoding
In **Manchester** and **differential Manchester** encoding, the clock pulse is embedded into the signal. Therefore, the receiver does not require any additional signal to represent the clock pulse. This self-clocking feature and low error rates have made Manchester and differential Manchester encoding the most popular line coding methods for wired LANs and WANs. According to the IEEE standards, Manchester encoding is used in Ethernet (IEEE 802.3) networks and differential Manchester encoding is used in Token Ring (IEEE 802.5) networks.

Fig. 2.18 NRZ encoding signals

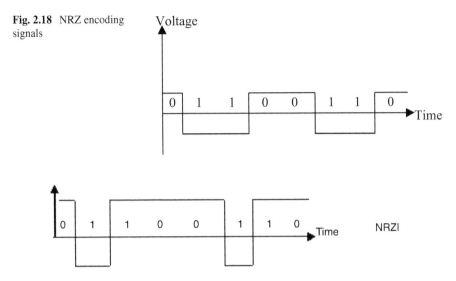

Fig. 2.19 NRZ-I encoding signals

Table 2.6 Conversion methods of digital signal to Manchester and differential Manchester

Digital signal	Manchester encoding	Differential Manchester
Logic 1	Transition from high to low at the middle of the signal	Transition only in the middle of the signal
Logic 0	Transition from low to high at the middle of signal	Transition at the start of zero and at the middle of zero (original signal)

Table 2.6 describes how to convert digital signals to Manchester encoding and differential Manchester encoding, and Fig. 2.20 shows the Manchester and differential Manchester encoding of a digital signal.

2.11 Error Detection Methods

When the transmitter sends a frame to the receiver, the frame can become corrupted due to external and internal noise. The receiver must first check the integrity of the frame. Some possible sources of error are as follows:

Impulse Noise: A non-continuous pulse for a short duration is called **impulse noise**. It may be caused by a lightning discharge or a spike generated by a power switch being turned off and on.

Crosstalk: This type of noise can be generated when a transmission line carrying a strong signal is coupled with a transmission line carrying a weak signal. The transmission line with the strong signal will produce noise (**crosstalk**) on the transmission line with the weak signal.

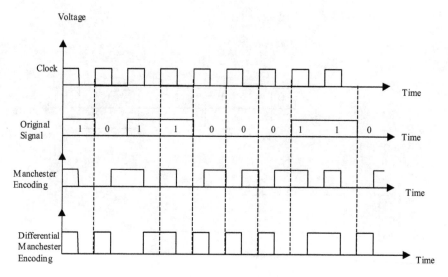

Fig. 2.20 Manchester and differential Manchester encoding

Attenuation: When a signal travels on a transmission line, the strength of the signal is reduced over distance. This reduction is called **attenuation**. A weak signal is more affected by noise than a strong signal.

White Noise or *Thermal Noise*: This type of noise exists in all electrical devices and is generated by moving electrons in the conductor.

The following methods can be used to detect an error or errors:

- Parity Check
- Block Check Character (BCC)
- One's Complement of the Sum
- Cyclic Redundancy Check (CRC)

Parity Check

The simplest error detection method is the **parity check**. The parity check method can detect one error and is used in both the asynchronous transmission method and the character-oriented synchronous transmission method. A parity bit is an extra bit that the transmitter adds to the information before transmitting to the receiver. The value of parity bit selected by the transmitter determines whether the data is given an even number of ones (even parity) or an odd number of ones (odd parity). For example, if a transmitter uses even parity to transmit the ASCII character 1000011 (upper case e), the transmitter adds parity bit 1 to the character so that the number of ones in the character becomes even: **1**1000011. The transmitter would then transmit 11000011 to the receiver. The receiver checks number of the ones in the character. If the number of the ones is even, there is no error detected in the character. Otherwise, the character contains an error. Parity error detection is used in serial communications. Figure 2.21 shows the logic diagram for a parity bit generator using Exclusive-OR gates.

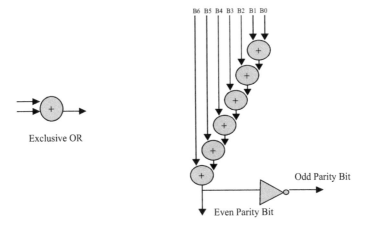

Fig. 2.21 Logic diagram of parity bit generator

Block Check Character

Block Check Character (BCC) uses vertical and horizontal parity bits in order to detect double errors. Remember, the parity check is limited to detection of only one error and is used only for transmitting single characters. When a block of characters is transmitted, BCC can be used to correct one error and detect two errors. A parity bit is added to each character (row parity) in a block of characters, then a column parity is computed. As illustrated in Table 2.7, the result of the column parity bits is called the Block Check Character (BCC). Table 2.7 shows that using odd parity for the rows and even parity for the columns results in a BCC of 01011110. For example, if two bits are changed in row one, such as B5 from 0 to 1 and B2 from 1 to 0, the row parity does not change, but the BCC will change, indicating the detection of two errors in one row.

One's Complement of the Sum

The **One's Complement of the Sum** method is used for error detection of the Transmission Control Protocol (TCP) header and Internet Protocol (IP) header. At the transmitter side, the 16 bit one's complement sum of the header is calculated. The result of this calculation is transmitted with the information to the receiver. At the receiver side, the 16 bit one's complement of the header is calculated and compared to the result with the one's complement of the transmitter. If the two results are equal, no error is detected. Otherwise, there is an error in the information. Figure 2.22 shows the one's complement of the sum for a four-byte header.

Cyclic Redundancy Check

The parity bit and BCC can detect single and double errors. The **Cyclic Redundancy Check (CRC)** method is used for detection of single error, more than a single error, and burst error (when two or more consecutive bits in a frame have changed). The CRC uses modulo-2 addition to compute the Frame Check Sequence (FCS). In modulo-2 addition:

Table 2.7 BCC calculation for word "NETWORK"

Parity	B6	B5	B4	B3	B2	B1	B0	
1	1	0/1	0	1	1/0	1	0	N
0	1	0	0	0	1	0	1	E
0	1	0	1	0	1	0	0	T
0	1	0	1	0	1	1	1	W
0	1	0	0	1	1	1	1	O
0	1	0	1	0	0	1	0	R
1	1	0	0	1	0	1	1	K
0	1	0/1	1	1	1/0	1	0	BCC

Transmitting Side	Receiving Side
1000010	1000010
1111101	1111101
0000001	0000001
+0111100	+ 0001100
1111100 (the carry is added)	1001100 (the carry is discarded)
Therefore: One's	Therefore: One's
Complement is 0000011	Complement is 0110011

Fig. 2.22 One's complement of the sum

$$1 + 1 = 0, 1 + 0 = 1, \text{and } 0 + 0 = 0.$$

The following procedure is used to calculate FCS.

At the transmitter side, the Frame M is k bits, P is a divisor of n + 1 bits, and the FCS is n bits and is equal to the remainder of 2^{n*} M/P using modulo-2 division.

At the transmitting side, the FCS, which is the remainder from 2^{n*} M/P, is calculated. The transmitter will transmit frame T = 2^{n*} M + FCS to the receiver, where T is k + n bits.

At the receiving side, the receiver divides T by P using modulo-2 division. If the result of this division generates a remainder of zero, no error is detected in the frame. Otherwise, the frame contains one or more errors.

Example 2.7 Find the Frame Check Sequence (FCS) for the following message. The divisor is given.

$$\text{Message } M = 111010, K = 6 \text{ bits}$$

$$\text{Divisor } P = 1101 \, n + 1 = 4 \text{ bits}$$

Therefore, $2^{n*}M = 111010000$.

By dividing $2^{n*}M$ by P using modulo-2 division, FCS = 010 as shown in Fig. 2.23.

At the transmitter side, the FCS is added to $2^{n*}M$, and the transmitter transmits frame T = 111010010 to the receiver.

Fig. 2.23 Frame Check
Sequence calculation

$$
\begin{array}{r}
101010 \text{ Quotient} \\
1101 \overline{)111010000} \\
\underline{1101} \\
1110 \\
\underline{1101} \\
1100 \\
\underline{1101} \\
\mathbf{010}
\end{array}
$$

Remainder of FCS is 010

At the receiver side, the receiver divides T by P, and if the result has a remainder of zero, there is no error in the frame. Otherwise, the message contains an error. Since the above division takes time, special hardware is designed to generate FCS.

CRC Polynomial and Architecture

A binary number is represented by $b_5b_4\,b_3b_2\,b_1b_0$, where b_i represents each bit that can be represented by a polynomial:

$$ b_5X^5 + b_4X^4 + b_3X^{3+}b_2X^2 + b_1X + b_0 $$

Example 2.8 Represent P = 1101101 by polynomial.

$$ P(X) = 1.X^6 + 1.X^5 + 0.X^4 + 1.X^3 + 1.X^2 + 0.X + 1 = X^6 + X^5 + X^3 + X^2 + +1 $$

The following CRC polynomials are IEEE and ITU standards:

$$ \text{CRC} \quad 12 \quad X^{12} \quad X^{11} \quad X^3 \quad X^1 \quad 1 $$

$$ \text{CRC} \quad 16 \quad X^{16} \quad X^{15} \quad X^2 \quad 1 $$

$$ \text{CRC} \quad \text{ITU} \quad X^{16} \quad X^{12} \quad X^5 \quad 1 $$

$$ \text{CRC} \quad 32 \quad X^{32} \quad X^{26} \quad X^{23} \quad X^{22} \quad X^{16} \quad X^{11} \quad X^{10} \quad X^8 \quad X^7 \quad X^5 \quad X^4 \quad X^2 \quad X \quad 1 $$

The CRC method uses a special integrated circuit (IC) to generate the FCS. The design of this IC is based on the CRC polynomial. In general, a CRC polynomial can be represented by:

$$ P(X) = X^n + ... + a_4X^4 + a_3X^3 + a_2X^2 + a_1X + 1 $$

Figure 2.24 shows the general architecture of a CRC integrated circuit (IC). Ci is a one-bit shift register and the output of each register is connected to the input of an Exclusive-OR gate; a_i is the coefficient of a CRC polynomial. In Fig. 2.24, if a_i equals zero, then there is no connection between the feedback line and the XOR

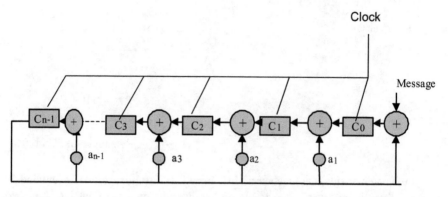

Fig. 2.24 General architecture of CRC polynomial

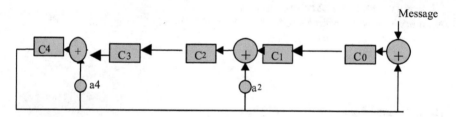

Fig. 2.25 CRC circuit for polynomial $P(X) = = X^5 + X^4 + X^2 + 1$

gate. In order to find the FCS, the initial value for Ci is set to zero, and the message $2^{n*}M$ is shifted $k + n$ times through the CRC circuit. The final content of $C_{n-1}, \dots$ $C_4\, C_3,\, C_2,\, C_1,\, C_0$ is the Frame Check Sequence (FCS).

Example 2.9
Show CRC circuit for polynomial:

$$P(X) == X^5 + X^4 + X^2 + 1$$

In the above polynomial the value for a1, a3 are zero and Fig. 2.25 shows the CRC circuit for above polynomial.

Example 2.10 Find FCS.

$$\text{Message M} = 111010$$

$$\text{Assume P} = 1101$$

$$P(X) = X^3 + X^2 + 1$$

The circuit for P(X) is shown in Fig. 2.26, where $a_1 = 0$, $a_2 = 1$, and $a_3 = 1$.

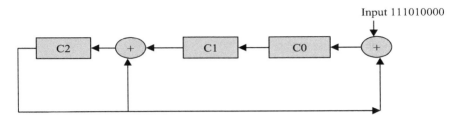

Fig. 2.26 CRC circuit for P = 1101

Table 2.8 FCS value for message M = 111010 and P = 1101

C2	C1	C0	Input
0	0	0	Initial value
0	0	1	1
0	1	1	1
1	1	1	1
0	1	1	0
1	1	1	1
0	1	1	0
1	1	0	0
0	0	1	0
0	1	0	0

Table 2.8 shows the contents of each register after shifting one bit at the time. After shifting 9 (k + n) times, the contents of the registers is the FCS.

Summary

- Information transfer between two computers occurs in one of two types of signals: digital or analog.
- Modern computers work with digital signals.
- A digital signal is represented by two voltages.
- Binary numbering is the representation of a number in Base-2.
- One digit in binary is called a bit, and eight bits are equal to one byte. More than one byte is called a word.
- Information is represented and processed inside the computer in binary or Base-2 form.
- Binary Coded Decimal (BCD) is used for representing decimal numbers from 0 to 9.
- ASCII code is used to represent character information inside the computer; ASCII code is made of 7 bits.
- There are two methods used for transmission of data: synchronous and asynchronous transmission.

- Parallel transmission is a method by which data is transmitted byte by byte.
- Serial transmission is a method by which data is transmitted one bit at a time over a single transmission media.
- Asynchronous transmission adds extra bits (start bit and stop) to the character for synchronization.
- With synchronous transmission, clock pulses are used for synchronization.
- There are three types of communication modes: Simplex transmission, Half-Duplex transmission, and Full-Duplex transmission.
- Baseband mode uses the bandwidth of a transmission media to carry only one signal.
- Broadband mode uses the bandwidth of a transmission media to carry multiple signals.
- Digital information can be represented by several forms of digital signal, such as non-return to zero (NRZ), non-return to zero inverted (NRZ-I), Manchester encoding, differential Manchester and bipolar encoding.
- Some sources of error for digital communications are impulse noise, crosstalk, attenuation, and white noise.
- Parity Check, Block Check Character (BCC), One's Complement of the Sum, and Cyclic Redundancy Checks (CRC) are used for error detection in networking.

Key Terms

Alternate Mark Inversion (AMI)	Full-Duplex Mode
Analog Signal	Half-Duplex Mode
ASCII	Manchester Encoding
Asynchronous Transmission	Non-return to Zero Encoding
Attenuation	One's Complement of the Sum
Baseband Mode	Parallel Transmission
Binary	Parity Check
Bipolar Encoding	Polar Encoding
Bit	Serial Transmission
Block Check Character (BCC)	Simplex Mode
Broadband Mode	Synchronous Transmission
Byte	Thermal Noise
Crosstalk	Unicode
Cyclic Redundancy Check	Unipolar Encoding
Differential Manchester Encoding	White Noise
Digital Signal	Word

Review Questions

Multiple Choice Questions

1. Frequency (F) is the number of cycles in one second and can be represented as:

   (a) $F = 1/T$
   (b) $F = T$
   (c) $F = -1/T$
   (d) $F = -T$

2. Modern computers work with _____ signals.

   (a) Digital
   (b) Analog
   (c) a and b
   (d) None of the above

3. Unicode is a new _____ bit character encoding standard code.

   (a) 16
   (b) 18
   (c) 8
   (d) 12

4. _____ transmission transmits data character by character.

   (a) Asynchronous
   (b) Synchronous
   (c) Full duplex
   (d) Half-duplex

5. _____ transmission uses asynchronous transmission.

   (a) Serial
   (b) Parallel
   (c) Broadband
   (d) Full duplex

6. In_____ mode, transmission of data goes only in one direction.

   (a) Simplex
   (b) Half duplex
   (c) Full duplex
   (d) Serial

7. In _____ mode, both computers can send and receive information simultaneously.

   (a) Simplex
   (b) Half duplex
   (c) Full duplex
   (d) Serial

8. The _____ of a communication signal is the range of frequencies that the signal occupies.

   (a) Data rate
   (b) Bandwidth
   (c) Baud rate
   (d) Broadband

9. What is the bandwidth of each computer for an Ethernet LAN with 20 computers? _____.

   (a) 1 Mbps
   (b) 10 Mbps
   (c) 500 Kbps
   (d) 2 Mbps

10. Cyclic Redundancy Check can_____.

   (a) Detect a single error and correct it
   (b) Detect double errors and correct them
   (c) Detect one or more
   (d) Correct more than one error

11. Which of following digital encodings carries clock pulse_____.

   (a) Manchester encoding
   (b) NRZ
   (c) RZ
   (d) RS-232

12. What is decimal value for $(111101)_2$?_____.

   (a) 44
   (b) 63
   (c) 61
   (d) 52

13. What is hexadecimal value for $(111110111)_2$?_____.

   (a) 1F6
   (b) 1F7
   (c) FB1
   (d) FB2

14. The binary value for 45 is _____.

   (a) 101011
   (b) 101101
   (c) 101111

(d) 011111

15. A range of frequencies carried by a medium is called.

   (a) Broadband signal
   (b) Baseband signal
   (c) Analog signal
   (d) A digital signal

16. Asynchronous communication uses:

   (a) Stop and start bits to indicate start of the character and end of the character
   (b) Start bit is used to synchronize transmission
   (c) Start and stop bits used for clocking
   (d) None of the above

17. What is the efficiency of serial connection using asynchronous transmission with 1 start bit, 2 stop bits, and 7 data bits?

   (a) 70%
   (b) 75%
   (c) 80%
   (d) 65%

Short Answer Questions

1. Sketch an analog signal.
2. What is frequency?
3. What is the unit of frequency?
4. What is the frequency of an analog signal that is repeated every 0.02 ms?
5. Explain the definition of the amplitude of an analog signal.
6. Sketch a digital signal.
7. What is a bit?
8. What is a byte?
9. What is a word?
10. Convert the following binary number to Hex.

$$(111000111001)_2 = (\underline{\hspace{2cm}})_{16}$$

11. Convert the following binary numbers to decimal.

$$(11111111)_2 = (\underline{\hspace{2cm}})_{10}$$

$$(10110001)_2 = (\underline{\hspace{2cm}})_{10}$$

12. Convert the following number to binary.

$$(FDE6)_{16} = (\underline{\hspace{2cm}})_2$$

13. Convert the word DIGITAL to binary using the ASCII table (Table 2.3).
14. Convert the word NETWORK to hexadecimal using the ASCII table (Table 2.3).
15. Write your name in binary ASCII, then change the result to hexadecimal.
16. What is serial transmission?
17. What is parallel transmission?
18. What is the advantage of parallel transmission over serial transmission.
19. Explain the following terms:

    (a) Simplex
    (b) Half Duplex
    (c) Full Duplex

20. What is a synchronous transmission?
21. Why is a clock pulse needed for transmission of a digital signal?
22. Show the format of asynchronous transmission.
23. Sketch a clock pulse.
24. List two types of digital encoding methods in which the clock is embedded to the data signal.
25. List methods of error detection.
26. List sources of error in networking.
27. Represent binary 110101 with a polynomial.
28. Find the BCC for word "ETHERNET."
29. Show the CRC Circuit for 1011.
30. Find the FCS for message 10110110 using circuit in question 29.
31. Find the One's Complement of the Sum for word "NETWORK."
32. Show the digital wave form for 0101011110.
33. Draw the Manchester encoding and differential Manchester encoding for the binary number 010110110.
34. Calculate the frequency of a signal repeated every 0.0005 seconds.
35. Find the FCS for data unit 111011 with divisor 1011.
36. What is burst error?

Chapter 3
Communications Channels and Media

Objectives

After completing this chapter, you should be able to:

- List the types of communication media currently in use.
- Distinguish between the different types of unshielded twisted-pair (UTP) cable.
- List the different types of coaxial cable and their applications.
- Discuss the different types of fiber-optic cables and their usage.
- Explain the operation of wireless transmission.
- Explain signal attenuation and channel bandwidth.
- Describe the characteristics of a synchronous optical network (SONET).
- List the components of SONET and define the function of each component.
- List SONET 's optical signal rates.
- Show the SONET frame format and explain the function of each overhead field.

3.1 Introduction

A transmission medium is a path between the transmitter and the receiver in a transmission system. The type of transmission medium is defined by the various characteristics of the digital signal, including the signal rate, data rate, and the bandwidth of a channel. The bandwidth of a channel determines the range of frequencies that the channel can transmit. There are three types of communications media currently in use:

1. Conductive, such as twisted-pair wire and coaxial cable.
2. Fiber-optic cable.
3. Wireless.

3.2   Conductive Media

The most popular **conductive media** used in networking are unshielded twisted-pair (UTP) cable, shielded twisted-pair cable (STP), and coaxial cable.

Twisted-Pair Cable

Unshielded twisted-pair (UTP) cable is the least expensive transmission medium and is typically used for LANs. Electrical interference, such as external electromagnetic noise generated by nearby cables, can have a devastating effect on the performance of a UTP cable. One way of improving the effect of noise on a UTP cable is to shield the cable with a metallic braid. A **shielded twisted-pair (STP) cable** provides better performance but is more difficult to work with. Figure 3.1a, b shows illustrations of UTP and STP cables.

The unshielded twisted-pair cable is divided into categories CAT-1 through CAT-8. Only CAT-1 through CAT-6, however, are recognized by the Electronic Industries Association (EIA) as CAT-7/7a and CAT-8 have not yet been standardized. Even so, CAT-7/7a cables are currently used in datacenters, and while CAT-8 is still largely in development, CAT-8 cables are projected for use in extremely high-bandwidth applications in the near future.

The EIA provides specifications for CAT-1 through CAT-6 UTP cables, as shown in Table 3.1. These standards apply to four-pair UTP which uses **RJ-45** and **RJ-11 connectors**, as shown in Fig. 3.2. In addition to these specifications, certain proprietary enhancements to CAT-5 allow for improved performance over longer distances. Table 3.1 also shows the specifications for the non-standardized CAT-7/7a and CAT-8 UTP cables.

Fig. 3.1 (**a**) Unshielded twisted-pair UTP) cable. (**b**) Shielded twisted-pair (STP) cable

Table 3.1 UTP Specifications

Type of UTP	Performance	Application
CAT-1	None	None
CAT-2	1 MHz	Telephone wiring
CAT-3	16 MHz	10 Base-T, 100 BaseT4
CAT-4	20 MHz	Token Ring 16
CAT-5/5e	100 MHz	100 Base-T, 1000 Base-T 20 Mbps Token Ring
CAT-6	250 MHz	10 Gbase-T (10 Gigabit Ethernet)
CAT-6a	500 MHz	10 Gbase-T (10 Gigabit Ethernet)
CAT-7	600 MHz	10 Gbps at 100 m
CAT-7a	1 GHz (1000 MHz)	40 Gbps at 50 m/100 Gbps at 15 m
CAT-8	2 GHz (2000 MHz)	25 Gbps/40 Gbps

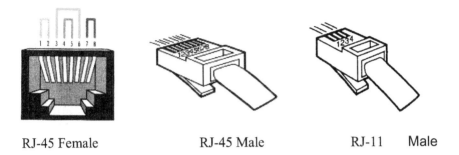

RJ-45 Female   RJ-45 Male   RJ-11   Male

Fig. 3.2 RJ-45 and RJ-11 connectors

Fig. 3.3 Coaxial cable

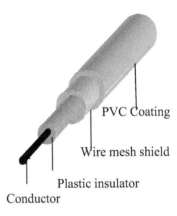

PVC Coating

Wire mesh shield

Plastic insulator

Conductor

Fig. 3.4 BNC connector

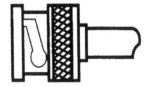

Coaxial Cable

A **coaxial cable** is used to transmit high-speed digital and analog signals over long distances. Figure 3.3 shows a coaxial cable that has an outer insulating cover made of polyvinyl chloride (PVC), or Teflon, protecting the coaxial cable. Under the outer cover is a wire mesh shield, which provides excellent protection from external electrical noise. This shield is made of mesh wire or foil, or both. Under the shield is a plastic insulator, which isolates the center conductor from the shield. The center conductor is a solid copper or aluminum wire that is shielded from external interference signals. There are different types of coaxial cable, categorized by the Radio Government (RG) rating. RG represents a set of specifications for cables such as the conductor diameter, thickness, and type of insulator. Coaxial cables use BNC connectors as shown in Fig. 3.4.

3.3   Fiber-Optic Cable

A **fiber-optic cable** is made of fiber that is covered by a buffer and a jacket. The fiber is composed of a core of thin glass or plastic covered by cladding which may also be glass or plastic. This fiber (the core and the cladding) is then covered by a buffer to strengthen it. The buffer is finally covered by a plastic outer layer, called the jacket, which acts as a protective coating or shield. Figure 3.5 illustrates the structure of a fiber-optic cable.

To transmit information using optical fiber, the digital information is converted to light pulses by **light-emitting diodes (LED)** or **injected-laser diodes (ILD)** and sent through the fiber-optic cable. An LED is a diode which generates a low power light. At the receiving end, a photodiode or a photo transistor is used to convert the light pulse signals back into electrical signals.

The following are the advantages of fiber-optic cables:

- Longer distance transmission due to reduced signal loss (attenuation).
- Greater bandwidth up to the Gigahertz range.
- Immunity from any kind of noise or external interference such as electromagnetic signals.
- Smaller size.
- Secure media.

Some disadvantages of fiber-optic cables are as follows:

- Network interface cards and cabling can be expensive.
- Connection to the network is more difficult.

Characteristics of Light
The source of signals for a fiber-optic cable is light. The characteristics of light are propagation speed, wavelength, and attenuation.

Propagation Speed   Light propagates through a vacuum at a speed of $3.0*10^8$ m/s.

Wavelength   The length of a wave is measured in meters and is represented by λ (lambda). The **wavelength** is the distance between two successive peaks of a wave or the distance traveled by one cycle of a wave as shown in Fig. 3.6.

Equation 3.1 describes the wavelength in terms of the speed of light and the frequency of a signal.

$$\lambda = C / f \tag{3.1}$$

Fig. 3.5   Fiber-optic cable

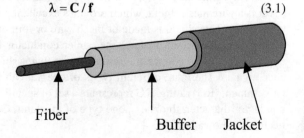

Fiber

Buffer   Jacket

voltage

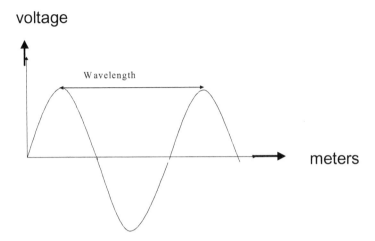

Fig. 3.6 Wavelength of a sine wave

where:

C = the speed of light ($3*10^8$ m/s)
f = frequency of the signal

Attenuation of Light Attenuation is the reduction of the strength of a signal. When light travels through a fiber, it loses energy. The greatest loss of energy is often caused by absorption. Absorption is caused by fiber materials as the optical power is converted to another form of energy such as heat. Attenuation is defined in Eq. 3.2.

$$A = 10\log_{10}\frac{P\,t}{Pr} \tag{3.2}$$

where:

A is the attenuation in decibels.
Pt is the power of light at the transmitter side.
Pr is the power of light at the receiver side (after transmission).

The attenuation of a fiber-optic cable is specified by the manufacturer. Figure 3.7 shows the attenuation of a 1 km fiber-optic cable with different signal wavelengths and two windows with the least attenuation (1300 nm and 1550 nm). The 850 nm wavelength window offers the most economical solution as it uses the less-expensive emitting diode. Fiber-optic systems operate at the wavelength defined by one of these three windows.

A. The first window is centered at a wavelength of 850 nm (nanometers = 10^{-9} meters).
B. The second window is centered at 1300 nm.
C. The third window is centered at 1550 nm.

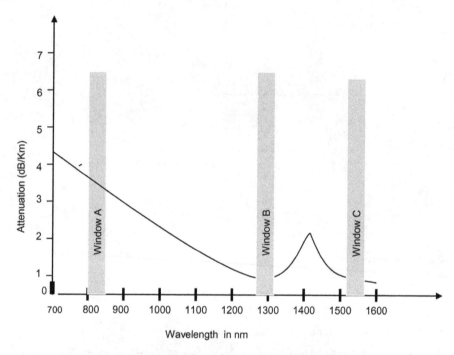

Fig. 3.7 Attenuation of fiber cable versus wavelength

Types of Fiber-Optic Cable

In a fiber-optic cable, the angle of light reflection is directly dependent upon the diameter of the fiber. As the diameter increases, the light is reflected more, and it takes more time to travel a given distance. There are two types of fiber-optics cable: single-mode fiber (SMF) and multimode fiber (MMF).

1. **Single-Mode Fiber (SMF):** In **single-mode fiber,** only one light ray propagates through the fiber as shown in Fig. 3.8. The core diameter of single-mode fiber is between 7 and 10 microns or micrometers (1 micron = 10^{-6} meters) and the cladding diameter of a single-mode fiber is 125 microns. The light wavelengths that are used for SMF are 1300 and 1550 nanometers. Manufacturers of fiber optic cables represent the fiber cable by ratio of core over cladding diameters such as 8/125 for single mode. The single mode fiber uses laser diode injection (LDI) as the source of light.

2. **Multimode Fiber (MMF):** In **multimode fiber,** more than one light ray can propagate through the fiber since each light ray propagates at a different wavelength. Multimode fiber has a core diameter larger than the wavelength of the light source being used, which ranges from 50 micrometers to 1000 micrometers, where the wavelength of the light is about 1 micrometer. This means light can propagate through the fiber in many different ray paths or modes. A single-

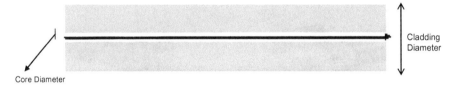

Fig. 3.8 Light propagation in single-mode fiber-optics (SMF)

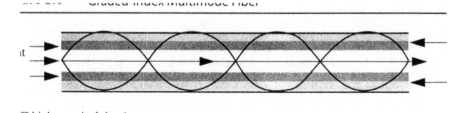

Fig. 3.9 Light propagation in multimode graded index fiber

mode fiber cable has a smaller diameter than a multimode fiber cable. The MMF uses a light emitting diode (LED) as the source of light.

Multimode Graded Index Fiber: In multimode fiber, the index of refraction across the core is gradually changed from maximum at the center to a minimum near the edges. This type of fiber causes the light to travel faster in the low index of refraction material than in the high-refraction material. Typical bandwidth for graded index fibers ranges from 100 MHz*km to 1 GHz*km. Figure 3.9 shows a multimode graded index fiber-optic cable.

Modal Bandwidth

Modal bandwidth is specified in units of MHz*km. The modal bandwidth indicates the amount of bandwidth supported by a fiber cable for a 1 km (0.625 miles) distance and is given by the manufacturer of the optical cable. For example, a cable with a modal bandwidth of 500 MHz*km can support end-to-end bandwidth of 250 MHz at a maximum 2 km (1.25 miles) distance.

Fiber-Optic Connectors

There are three common types of fiber-optic connectors used for networking. These connectors are listed below.

1. **Subscriber Channel (SC) Connector**: The SC connector, shown in Fig. 3.10, uses a push-pull locking system. SC connectors are used for CATV, telephone connections, and networks.
2. **Straight Tip (ST) Connector**: ST connectors use bayonet locking and are valued for their high reliability. The ST connector is also shown in Fig. 3.10.
3. **MT-RJ Connector**: The MT-RJ is a duplex connector, as shown in Fig. 3.11. The size of a MT-RJ connector is equal to that of an RJ-45 connector.

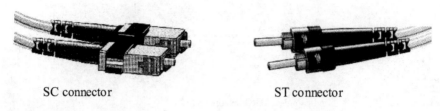

SC connector   ST connector

Fig. 3.10 Fiber-optic ST and SC connectors

Fig. 3.11 MT-RJ
connector

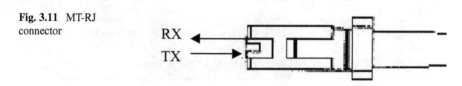

RX

TX

3.4 Wireless Transmission

Wireless transmission does not use any transmission media, such as a conductor or optical cable, to transmit and receive information. Microwave, radio, infrared light, and laser are forms of wireless communication.

When electrons accelerate, they generate electromagnetic waves. Wireless transmission uses these electromagnetic waves. Table 3.2 shows the electromagnetic wave spectrum and its applications.

3.5 Transmission Impairment

In a communication system, the transmitter sends information in the form of signals such as optical, electrical, or radio frequency to the receiver. The signals are sent on a communication channel. The receiver side of the signal has a different shape from the transmitter side. This difference can be caused by attenuation, distortion, or noise.

Attenuation When a signal travels from source to destination, the signal loses energy. The amount of lost energy depends on the type of the channel, the length of the channel, and the frequency of the signal, as shown in Fig. 3.12. A longer channel will result in a higher loss of energy than a shorter channel. If a signal attenuates too much, then it becomes undetectable by the receiver.

To increase the length of a communication channel, one must use an amplifier to strengthen the signal. The attenuation is measured in decibels (dB) and it is defined by Eq. 3.3.

Table 3.2 The electromagnetic frequency spectrum and its applications

Frequency range	Name	Application
3–30 KHz	Very low frequency (VLF)	Telephone
30–300 KHz	Low frequency (LF)	Radio frequency for navigation
300–3000 KHz	Medium frequency (MF)	AM radio frequency
3–30 MHz	High frequency (HF)	CB radio, shortwave radio
30–300 MHz	Very high frequency (VHF)	TV and FM radio
300–3000 MHz	Ultra high frequency (UHF)	TV, military
3–30 GHz	Super high frequency (SHF)	Terrestrial and satellite microwave
30–300 GHz	Extreme high frequency (EHF)	Experimental
>300 GHz	Infrared light, lasers	TV remote control, laser surgery

Fig. 3.12 Attenuation and amplification of a signal as it travels through a communication channel

$$A_p = 10\log_{10} \frac{Pt}{Pr} \qquad (3.3)$$

where:

A_p is the power attenuation.
Pt is the power of the signal at the transmitter side.
Pr is the power of the signal at the receiver side.

Example 3.1 A signal with a power of 500 mW is transmitted over a communication channel. At the receiver side, the power of the signal is 50 mW. Calculate power attenuation of the signal.

$$A_p = 10\log_{10} \frac{500}{500} = 10dB$$

Electrical signals are used for transmission of information over a conductive medium that acts as a communication channel. The electrical signal travels through the conductor. As a result of the resistance of the conductor, the signal loses some voltage. This loss of voltage is called a voltage drop. Attenuation of an electrical signal is calculated by Eq. 3.4:

$$Pt = Vt^* I, \text{where } I = Vt / R, \text{then}$$

$$Pt = \left(Vt\right)^2 / R$$

$$Pr = \left(Vr\right)^2 / R$$

Substitution of Pt with $(Vt)^2 / R$ and Pr with $(Vr)^2 / R$ results in Eq. 3.4

$$Av = 20\log_{10}\frac{V_t}{V_r} \tag{3.4}$$

where:

Av is the voltage attenuation.
V_t is the voltage of signal at the transmitter side.
V_r is the voltage of signal at the receiver side (after transmission).

The attenuation of a cable is published by the cable manufacturer. Table 3.3 shows the attenuation of UTP Cat-5 and Cat-6 cables at different frequencies. It is notable that attenuation increases as frequency increases.

The network designer uses attenuation data to find the maximum cable length that can be used without using a repeater.

Example 3.2 Find the maximum length of a Cat-6 cable that transmits a signal with 250 MHz, assuming the voltage of the signal at the transmitter side is 5000 mV and at the receiver side is 200 mV.

$$A_v = 20 \ \log_{10}\frac{5000}{200} = 27.8dB$$

From Table 3.3, attenuation for 100 meters of Cat-6 at frequency 250 MHz is 31 dB; thus, the maximum cable length is 89.6 meters. If the transmitter transmits information at 20 MHz using Cat-6, then the maximum length of the cable would be 347.4 meters.

Table 3.3 Attenuation of Cat-5 and Cat-6 cables for various frequencies

	Cat-5	Cat-6
Frequency (MHz)	Attenuation (dB/100 m)	Attenuation (dB/100 m)
1 MHz	2	1.9
10	6.5	5.6
20	9.3	8.0
100	22	18.7
250	NA	31.0

3.6   Bandwidth, Latency, Throughput, and Channel Capacity

Bandwidth
In general, bandwidth is the maximum rate of data transferred over a communication link. It is categorized by the type of signal over the communication link.

Analog Bandwidth   Analog bandwidth is the difference between the highest and the lowest frequency in a communication channel. For example, the highest frequency of the human voice is $3300\,Hz$ and the lowest frequency of the human voice is $300\,Hz$. Therefore, the bandwidth of the human voice is simply:

$$3300\,Hz - 300\,Hz = 3000\,Hz$$

Digital Bandwidth   The bandwidth of a digital link is the maximum number of bits per second that can be transmitted over the communication link. For example, the bandwidth of a T1 link is 1.54 Mbps meaning that it can transfer up to 1.54 million bits per second. Ethernet's bandwidth is 10 Mbps meaning each bit takes 0.1 µs to get transmitted.

Latency (Delay)
Latency defines the time it takes to transmit one packet (unit of information) from source to destination. Latency delay consists of propagation delay, transmission time, and buffering time. Latency is defined in Eq. 3.5. Also, the two-way latency is called round trip time (RTT).

$$\text{Latency} = T_x + T_p + T_b \tag{3.5}$$

where:

T_x is the transmission times.
T_p is the propagation delays.
T_b is the buffering times.

Transmission Time   Transmission time is the time that it takes to put a message on media. If the data rate of a link is 1000 bits per second, then each bit takes 0.001 seconds to put on the media. Transmission time is defined by Eq. 3.6:

$$\text{Transmission Time } T_x = \frac{\text{Packet Size}\,(\text{bits})}{\text{Bandwidth}\,(bps)} \tag{3.6}$$

Example 3.3   Find the transmission time for transferring 1500 bytes using a communication link with a data rate of 10 Mbps.

$$T_x = \frac{1500(byte) \times \left(8 \dfrac{bit}{byte}\right)}{10 \times 10^6 \dfrac{bit}{sec}} = 0.0012 \text{ sec}$$

Propagation Delay Propagation delay or propagation time is the time taken by the signal to travel from source to destination. Propagation delay is defined by Eq. 3.7.

$$T_p = \frac{\text{Length of the Communication Link} (\text{meters})}{\text{Speed of Light in the medium}} \tag{3.7}$$

where:

Speed of light = $3 \times 10^8 \, m/s$ in a vacuum and less in wire and fiber medium.

Electrical and optical signals travel considerably less than almost at the speed of light and are generally taken as $2.3 \times 10^8 \, m/s$ and $2 \times 10^8 \, m/s$, respectively.

Example 3.4 Find the propagation time for transferring 100 bytes over 200 km of fiber-optic cable.

$$T_p = \frac{2 \times 10^5 \, m}{2 \times 10^8 \left(\dfrac{m}{s}\right)} = 0.001 \text{sec}$$

Buffering Time A transmitted packet might be stored in several locations before reaching its final destination. The time a packet spends in a temporary location (known as a buffer) is called buffering time or queue time.

Throughput
The throughput of a communication channel is defined by the number of bits transmitted over a communication link and can be shown by Eq. 3.8.

$$\text{Throughput} = \frac{\text{Transfer Size}}{\text{Latency}} \tag{3.8}$$

Example 3.5 Calculate the transmission time and the throughput of a communication link for a user to download 1500 bytes of information from a server. The user's computer is connected to the server by a modem with the data rate of 50 Kbps and the distance between the two computers is 4000 km. Assume there is no buffering delay.

$$T_x = \frac{1500 \text{ byte} \times 8 \left(\dfrac{\text{bit}}{\text{byte}} \right)}{50,000 \text{ bps}} = 0.24 \text{ sec}$$

$$T_p = \frac{4 \times 10^6 \text{ m}}{2.3 \times 10^8 \left(\dfrac{\text{m}}{\text{s}} \right)} = 0.018 \text{ sec}$$

$$\text{Latency} = 0.24 + 0.018 = 0.258 \text{ sec}$$

$$\text{Throughput} = \frac{1500 \text{ byte} \times 8 \left(\dfrac{\text{bit}}{\text{byte}} \right)}{0.258 \text{ sec}} = 46.51 \text{ kbps}$$

Channel Capacity

The bandwidth of a channel is defined as the range of frequencies that pass through the channel. Nyquist's theorem is defined as the maximum data rate $\left(\dfrac{\text{bits}}{\text{sec}} \right)$ in a noiseless channel and is represented mathematically by Eq. 3.9.

$$\text{max Data Rate} \left(\text{MDR} \right) = 2W^* \log_2 N \ bps \qquad (3.9)$$

where:

W is the bandwidth of the channel.
N is the number of signal levels or voltage levels.

Example 3.6 Find the maximum data rate of a channel with a bandwidth of $4000 \, Hz$ transmitting two voltages (e.g., binary = two levels, 0 and 1).

$$MDR = 2^* 4000^* \log_2 2$$
$$MDR = 8000 \ bps$$

Equation 3.9 is valid only when using a noiseless channel. Noise affects the data rate of a channel. Figure 3.13 shows transmission of a digital signal through a noisy channel.

Figure 3.13 indicates that the presence of noise in a communication channel may cause distortion of the incoming signal such that the output signal is no longer a square wave. When the amplitude of the noise is larger than expected, total destruction of the original signal occurs and affects the data rate of the communication channel. The capacity of a channel may be obtained using Shannon's theorem and is represented mathematically by Eq. 3.10.

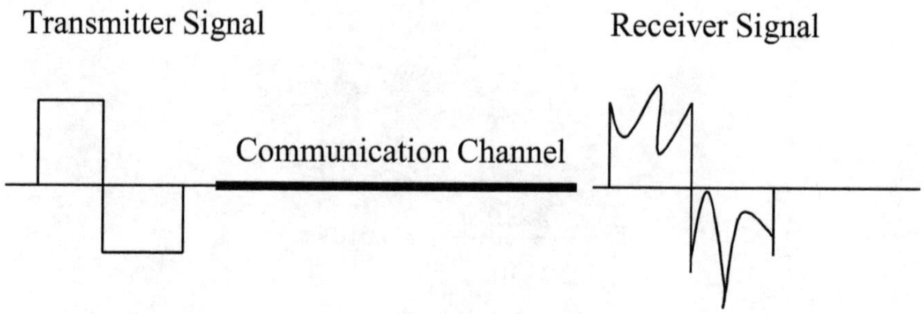

Transmitter Signal   Receiver Signal

Fig. 3.13 Transmission of a digital signal through a noisy channel

$$\text{max Data Rate}\,(\text{MDR}) = W^* \log_2\left(1+\frac{S}{N}\right)\text{bps} \qquad (3.10)$$

where:

$\frac{S}{N}$ is the signal-to-noise ratio(SNR), the ratio of average signal power to average noise power at the receiver, which is usually given in decibels(dB)by Eq. 3.11.

A decibel is defined as:

$$Decibels = 10^* \log_{10}\left(\frac{S}{N}\right) \qquad (3.11)$$

For example, if $\left(\frac{S}{N}\right) = 10$, then ($SNR$) in decibels is:

$$SNR = 10^* \log_{10} 10$$
$$= 10\,dB$$

Example 3.7 Find the maximum data rate of a channel with a bandwidth of $4000\,Hz$ and a signal-to-noise ratio of $30\,dB$.

$$30dB = 10^* \log_{10}\frac{S}{N}$$

Therefore,

$$\frac{S}{N} = 10^3$$

Subsequently,

$$MDR = W^* \log_2\left(1+10^3\right)$$

$$= 4000^{*}9.967 bps$$

3.7 Synchronous Optical Network (SONET)

SONET is a high-speed optical carrier that uses fiber optic-cable as the transmission media. The term SONET is used in North America and is a standard established by the American National Standards Institute (ANSI). The International Telecommunication Union (ITU) has also set a standard for SONET called Synchronous Digital Hierarchy (SDH), which is used in Europe.

SONET optical architecture is based on a four-fiber bidirectional ring to provide the highest possible level of service assurance. New application software, such as Medical Images and CAD CAM applications, require more bandwidth than other applications and rely on SONET for high-speed transmission with a large bandwidth.

3.7.1 Characteristics of SONET

The most significant characteristics of SONET are as follows:

- SONET uses byte multiplexing at all levels.
- SONET is a high-speed transport (carrier) technology with a self-correcting path.
- SONET uses multiplexing and demultiplexing.
- SONET provides **operation administration and maintenance (OAM)** functions for network managers.
- The basic electrical signal for SONET is s**ynchronous transport signal level one (STS-1).**
- SONET transmits STS-1 at a rate of 8000 frames per second.
- Slower signals can be multiplexed directly onto higher speeds.

3.7.2 SONET Components

Figure 3.14 shows SONET's components, which consist of an STS (synchronous transport signal) multiplexer (MUX), a regenerator, an add/drop multiplexer, an electrical-to-optical converter, and an STS demultiplexer. These components are described in the following list:

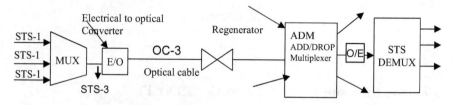

Fig. 3.14 SONET components

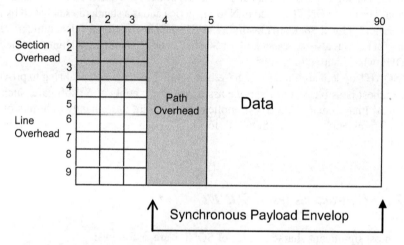

Fig. 3.15 Multiplexing three STS-1 s to generate STS-3

STS Multiplexer: The function of the **STS MUX** is to multiplex electrical input signals to a higher data rate and then convert the results to an optical signal, as shown in Fig. 3.15.

Regenerator: The regenerator performs the functions of a repeater. If the optical cable is longer than standard, the regenerator will be used to receive the optical signal and then regenerate the optical signal.

Add/Drop Multiplexer: Add/drop multiplexers are used for extracting, or inserting, lower rate signals from, or into, higher rate multiplexed signals without completely demultiplexing the SONET signals.

STS Demultiplexer: STS demultiplexers convert and demultiplex optical signals to electrical signals.

3.7.3 SONET Signal Rates

The lowest level signal in SONET is the Synchronous Transport Signal Level One (STS-1), which has a signal rate of 51.84 Mbps. The STS-1 is an electrical signal which is converted to an optical signal called OC-1. The higher SONET data rates are represented by STS-n, where n is 1, 3, 9, 12, 18, 24, 34, 48, 96, and 192. Table 3.4 shows SONET and SDH signal rates.

Table 3.4 Data rate for OC, STS, and STM signals

Fiber-Optic (OC) signal OC-n Level	Synchronous transport signal (STS) for SONET	Synchronous transport module data rate (STM) for SDH (Mbps)	
OC-1	STS-1	51.84	
OC-3	STS-3	STM-1155.52	
OC-9	STS-9	STM-3446.56	
OC-12	STS-12	STM-4622.08	
OC-18	STS-18	STM-6933.12	
OC-24	STS-24	STM-8	1244.16
OC-36	STS- 36	STM-12	1866.24
OC-48	STS-48	STM-16	2488.32
OC-96	STS-96	STM-32	4976.64
OC-192	STS-192	STM −64	9953.28

OC Optical carrier, *STS* Synchronous transport signal (electrical signal for SONET), *STM* Synchronous transport module (electrical signal for SDH)

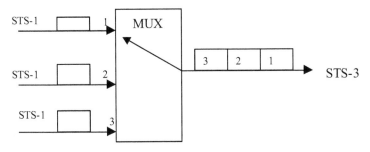

Fig. 3.16 SONET STS-1 frame format

3.7.4 SONET Frame Format

The basic transmission signal for SONET is the STS-1. The STS-1 format is shown in Fig. 3.16. It is made up of 9 rows and 90 columns of bytes. The frame size is 90 * 9 = 810 bytes, or 810 * 8 = 6480 bits. SONET transmits 8000 frames per second. The data rate for STS-1 is 6480 * 8000 = 51.84 Mbps.

The first three columns are referred to as transport overhead, which is 3 * 9 = 27 bytes. 9 of these 27 bytes are used for section overhead, 18 bytes are used for line overhead, and 9 bytes are used for path overhead. The actual data rate is 86 columns * 9 rows * 8 bits * 8000 frames/sec = 49.536 Mbps.

The STS-1 frame is transmitted by the byte from row 1, column 1 to row 9, column 90 (scanning from left to right).

Path Overhead Path overhead is part of SPE (synchronous payload envelope) and contains the performance monitor of synchronous transport signal, path trace, parity check, and the path status.

Section Overhead Section overhead contains information about frame synchronization.

(informing the destination of incoming frame) and frame identification. It also carries information about operation administration and maintenance (OAM), handles frame alignment, and separates data from the voice.

Line Overhead Line overhead carries the payload pointers to specify the location of SPE in the frame and provides automatic switching for standby equipment. It separates voice channels and provides multiplexing, line maintenance, and performance monitoring.

3.7.5 SONET Multiplexing

Higher levels of synchronous transport signals can be generated by using byte interleave multiplexing. The STS-3 is generated by multiplexing three STS-1 signals as shown in Fig. 3.15. The output of STS-3 is converted to an optical signal called OC-3. The STS-3 frame is made up of 3 * 90 = 270 columns, and 9 rows, containing 2430 bytes. The STS-3 is transmitted at 8000 frames per second and, therefore, the bit rate of STS-3 is:

$$2430 \, \text{bytes}^* \, 8 \, \text{bits}^* \, 8000 \, \text{frames} / \sec = 155.52 \, \text{Mbps}.$$

Figure 3.17 shows the STS-3 frame format. The transport overhead is made up of 9 columns and 9 rows. The SONET payload envelope is 260 * 9 bytes. The STS-9 is generated by multiplexing three STS-3 s, as shown in Fig. 3.18.

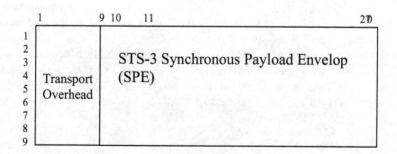

Fig. 3.17 Frame format of STS-3

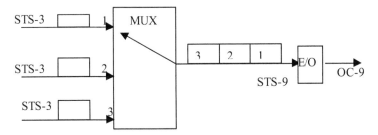

Fig. 3.18 Generating an STS-9 frame

Summary

- Transmission media are used to connect computers.
- The types of communication media are conductors, optical cable, and wireless.
- The types of conductors used for networking are unshielded twisted pair (UTP) cable, shielded twisted pair (STP) cable, and coaxial cable.
- UTP cable contains several pairs of wires and is divided by EIA into categories such as Cat-1, Cat-5, and Cat-6.
- Unshielded twisted pair (UTP) uses an RJ-45 connector.
- A coaxial cable is used for transmitting high-speed information over relatively long distances.
- A fiber-optic cable transfers information in the form of light.
- Light emitting diodes (LED) and laser diodes are used to convert a digital signal to an optical signal for transmitting information over optical cable.
- There are two types of optical cable: single-mode fiber (SMF) and multimode fiber (MMF).
- Single-mode fiber (SMF) only uses one ray of the light source.
- In MMF-grade index fiber, the index of refraction changes from maximum at the center to minimum near the edge of the core, causing the light to bend in a curved shape.
- Wireless transmission uses microwave, radio, or infrared light signals to transmit information.
- Microwave communication uses radio waves in the range of 1 GHz to 23GHz.
- Channel bandwidth is defined as the range of frequencies that passes through the channel.
- Analog bandwidth is the difference between the highest and the lowest frequency in a communication channel.
- Digital bandwidth is the maximum number of bits per second that can be transmitted over a communication link.
- Latency (delay) defines the time it takes to transmit one packet (unit of information) from source to destination.
- Attenuation is the loss of energy in a signal after transmission.

- The synchronous optical network (SONET) is a high-speed data carrier. The term SONET is used in North America and Synchronous Digital Hierarchy (SDH) is used in Europe.
- SONET uses fiber-optic cable for transmission media since optical transmission is immune to interference and can transmit data over a long distance.
- SONET converts the synchronous transport signal (STS-1) to an optical signal. It is called OC-1 (optical carrier) and is transmitted at a rate of 8000 frames per second.
- SONET components are the STS multiplexer, the regenerator, the add/drop multiplexer (ADM), and the STS demultiplexer.
- The basic transmission signal for SONET is the STS-1. The frame format of SONET is made of 9 rows and 90 columns of bytes. The frame size is 810 bytes and this frame is transmitted at a rate of 8000 frames per second, which gives a bit rate of 51.84 Mbps for STS-1.
- Three STS-1 signals can be multiplexed and converted to an optical signal to generate an OC-3 with a data rate of 155.52 Mbps.

Key Terms

Add/Drop Multiplexer	Optical Carrier Signal (OC)
Attenuation	Path Overhead
Bandwidth	Propagation Delay
Channel Capacity	Regenerator
Coaxial Cable	RJ-45 and RJ-11 Connectors
Conductive Media	Section Overhead
Decibel	Shielded Twisted pair
Fiber-Optic Cable	Single Mode Fiber (SMF)
Injected-Laser Diode (ILD)	SONET Frame Format
Latency	STS Demultiplexer
Light-Emitting Diode (LED)	STS Multiplexer
Line Overhead	Synchronous Optical Network (SONET)
Microwave	Synchronous Payload Signal (SPE)
Modal Bandwidth	Synchronous Transport Signal (STS)
MT-RJ Connector	Synchronous Transport Signal Level One (STS-1)
Multimode Fiber (MMF)	Transmission Time
Multimode Graded Index Fiber	Unshielded Twisted Pair (UTP)
Nyquist Theorem	Wavelength
Operation Administration and Maintenance (OAM)	Wireless Transmission

Review Questions

Multiple Choice Questions

1. A _____ cable is the least expensive transmission media.

   (a) UTP
   (b) STP
   (c) Fiber-optic
   (d) Coaxial

2. _____ cable is used to transmit high-speed and analog signals.

   (a) UTP and coaxial
   (b) STP and UTP
   (c) Coaxial and fiber-optic
   (d) Fiber-optic and STP

3. _____ connector(s) is/are used in fiber-optic cable.

   (a) SC and BNC
   (b) ST and SC
   (c) RJ-11 and ST
   (d) BNC and SC

4. _____ transmission does not use any transmission medium.

   (a) WAN
   (b) LAN
   (c) Wireless
   (d) Internet

5. Wireless transmission uses _____ waves.

   (a) Optical true
   (b) Electrical
   (c) Electromagnetic
   (d) Digital

6. Which of the following UTP cables is suitable for a data rate of 100 Mbps?

   (a) Cat-2
   (b) Cat-4
   (c) Cat-3
   (d) Cat-5

7. Which of the following transmission media is used for high-speed transmission?

   (a) Coaxial cable and fiber-optic cable
   (b) Fiber-optic cable and UTP CAT-2 cable
   (c) Microwave and fiber-optic cable
   (d) UTP CAT-2 cable and microwave cable

8. What type of fiber-optic cable is used for long distance transmission?

   (a) Multimode graded index
   (b) Single-mode
   (c) UTP
   (d) STP

9. What is the speed of light in a vacuum?

   (a) $3*10^8$ m/s
   (b) $5*10^8$ m/s
   (c) $3*10^2$ m/s
   (d) $6*10^{20}$ m/s

10. What is the maximum length that a fiber cable with a modal bandwidth of 1000 MHz*km can be in order to transmit information with a 200 MHz speed?

    (a) 1 km
    (b) 10 km
    (c) 5 km
    (d) 20 km.

11. _____ sets the standard for SONET.

    (a) IEEE
    (b) ANSI
    (c) ITU
    (d) ISO

12. SONET uses byte multiplexing in _____ levels.

    (a) Upper
    (b) Mid
    (c) All
    (d) None of the above

13. The basic electrical signal for SONET is _____

    (a) STS-1
    (b) STS-2
    (c) STS-3
    (d) STS-n

14. SONET transmits the STS-1 at a rate of _____ frames/second

    (a) 6000
    (b) 7000
    (c) 8000
    (d) 1000

15. The _____ performs the function of a repeater.

   (a) Regenerator
   (b) STS multiplexer
   (c) STS demultiplexer
   (d) SONET

16. STS-1 has data rate of _____ Mbps.

   (a) 810
   (b) 8000
   (c) 51.84
   (d) 1.54.

17. SONET is a/an_____

   (a) LAN
   (b) WAN
   (c) Optical carrier
   (d) Internet

18. An STS-1 frame is made of_____.

   (a) 9 columns and 90 rows
   (b) 9 rows and 90 columns
   (c) 10 rows and 100 columns
   (d) None of the above.

19. The optical signal for STS-1 is _____.

   (a) OC-3
   (b) OC-1
   (c) OC-2
   (d) OC-n

20. An STS-3 is generated by multiplexing_____.

   (a) Three STS-1s
   (b) Six STS-1s
   (c) Five STS-1s
   (d) Two STS-1s

21. An STS-9 is generated by multiplexing.

   (a) 6 STS-1s
   (b) 3 STS-3s
   (c) 3 STS-1s
   (d) Two STS-3s

22. An STS-3 frame format is made up of _____

(a) 270 rows and 9 columns
(b) 9 rows and 270 columns
(c) 10 rows and 300 columns
(d) None of the above

Short Answer Questions

1. List the major communications media.
2. What does UTP stand for?
3. What does STP stand for?
4. Which organization defines standards for cables?
5. What is the performance of a Cat-5 UTP cable?
6. What type of light source is used for fiber-optic cables?
7. What are the advantages of fiber-optic cables over conductive cables?
8. What are the types of fiber-optic cables?
9. What does SMF stand for?
10. What does MMF stand for?
11. What is the application of single-mode fiber?
12. What is the application of multimode fiber?
13. What are the types of microwave communication?
14. What is the range of microwave frequencies?
15. What are the signal sources for optical cabling?
16. What are the advantages of optical cables over coaxial cables?
17. What are the advantages of STP cables over UTP cables?
18. Explain digital bandwidth.
19. Explain latency and the causes of latency.
20. Find the transmission time of a packet of 1500 bytes transmitted over a 100 Mbps channel.
21. For the following cases, what should the transmitter voltage be for 100 meters of Cat-5 cable with a receiver voltage of 500 mV?

(a) Transmitted signal at 20 MHz
(b) Transmitted signal at 100 MHz

22. 2000 bytes of data are to be transferred between a server and a host computer, which are connected via a 1000-meter Cat-5 cable with a transmission rate of 10 Mbps. Calculate the following:

(a) Transmission time
(b) Propagation delay
(c) Round trip time

23. A packet of 100 bytes is transmitted over a 100 km cable with bandwidth of 100 Mbps. Calculate the following:

(a) Propagation delay of the link
(b) Transmission time
(c) Latency of the packet
(d) RTT.

24. What is the bandwidth of a 20 km link for transmitting 500 bytes of information such that the propagation delay is equal to transmission delay?

25. Find the time that it takes to transmit 1000 kByte-files from a server that is located 4000 km away from a host computer. Assume you are using a modem with a data rate of 52 Kbps and size of each packet is 1000 bytes.

26. Calculate the latency for transmitting 1500 bytes of data over the following links:

(a) 100 meters copper with a bandwidth of 10 Mbps
(b) 4000 meters optical fiber with a bandwidth of 10 Mbps

27. 500 bytes of data are transmitted over 200 km of a fiber-optic cable.

(a) Find the data rate such that the transmission time becomes equal to the propagation time.
(b) What is throughput of this communication link?

28. Find the maximum data rate of a communication link with a bandwith of $3000\,Hz$ using 8 signal levels.

29. Find the bandwidth of communication channel in order to transfer data at a rate of 100 Mbps, assume $\dfrac{S}{N}$ ratio is($50\,dB$).

30. What does SONET stand for?
31. What does SDH stand for?
32. What is an application of SONET?
33. What is the basic electrical signal for SONET?
34. What is the transmission media for SONET?
35. List some of the advantages of SONET.
36. List the SONET components.
37. What does STS-1 stand for?
38. What is OC-1?
39. What is the data rate for STS-1?
40. How many bytes is STS-1?
41. How many STS-1 must be multiplexed to generate an STS-3?
42. SONET transmits how many frames per second?
43. Show the SONET frame format.
44. Explain the function of add/drop multiplexing.
45. What is STS-n?
46. Why is the STS-1 bit rate 51.84 Mbps?

Chapter 4
Multiplexer and Switching Concepts

Objectives
After completing this chapter, you should be able to:

- Explain the operation of multiplexers and demultiplexers.
- List the types of multiplexers.
- Discuss how a telephone system operates.
- Explain how pulse code modulation converts voice to digital signals.
- Explain T1 Link technology and how to calculate its data rate.
- Discuss switching concepts.
- List the types of switching methods.

4.1 Introduction

Long-distance transmission lines are expensive, so a method that allows several devices to share one transmission line is necessary to defray the cost of wiring. The multiplexer provides a solution to this problem. Figure 4.1 shows four terminals sharing one transmission line to send information to the host computer by using a multiplexer instead of four transmission lines.

A **multiplexer (MUX)** is a device that combines several low-speed data channels and transmits all the data on a single high-speed channel. A common application of multiplexing is long-distance communication using high-speed point-to-point links for transferring large quantities of voice signals and data between users. Figure 4.2 shows the basic architecture of a multiplexer. A multiplexer that has N inputs and one output is called a N-to-1 multiplexer. Figure 4.2 shows a 4-to-1 Multiplexer. The internal switch selects one input line at a time and transfers the input to the output. When the switch is in position A, it transfers input A to the output. Then, the switch moves to position B and transfers input B to the output.

A. Elahi, A. Cushman, *Computer Networks*,
https://doi.org/10.1007/978-3-031-42018-4_4

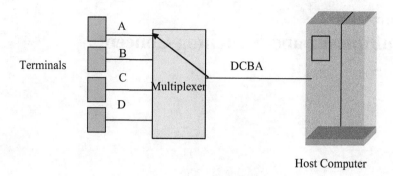

Fig. 4.1 Application of multiplexer

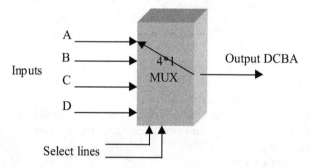

Fig. 4.2 4-to-1 multiplexer

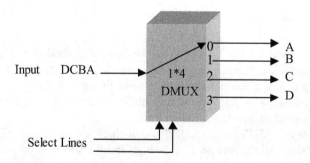

Fig. 4.3 1-to-4 demultiplexer

This method continues until the switch moves to position D and transfers input D to the output. After this function is completed, the switch starts over from input A.

The opposite of a multiplexer is a **demultiplexer (DMUX)**, as shown in Fig. 4.3. The switch moves to send each input to the appropriate output. A DMUX has one input and N outputs – this is called a 1-to-N demultiplexer. When the switch is in position 0, it transfers A to output port 0, then moves to output port 1 and transfers B to this port. This process continues until the switch moves to output port 3 and transfers D to port 3. Once the cycle is complete, the transfer of data starts over from port 0.

4.2   Types of Multiplexers

Multiplexers are categorized into the following types, where each type has a specific application:

1. Time division multiplexing (**TDM**)
2. Frequency division multiplexing (**FDM**)
3. Statistical packet multiplexing (**SPM**)
4. Fast packet multiplexing (**FPM**)
5. Code division multiplexing (**CDM**)
6. Wavelength division multiplexing (**WDM**)

Time Division Multiplexing (TDM)
In **time division multiplexing (TDM)**, multiple digital signals can be carried on a single transmission path by interleaving each input of the multiplexer. A TDM operates with a preassigned equal time slot to each input. It divides the bandwidth of the multiplexer's output into fixed segments, and each input to the MUX is given a fixed unit of time. First, information from input one is transmitted, then from input number two, and so on in a regular sequence, as shown in Fig. 4.4.

One disadvantage of TDM is that the bandwidth of any one input is not available to other inputs when that input to the TDM is inactive. In Fig. 4.4, the inputs to B and D are inactive at times t1 and t2, respectively, and thus the outputs in frame #2 and #3 have idle times. Another disadvantage of TDM is that it is not able to change the bandwidth of the input dynamically and, therefore, cannot transport a combination of voice, fax, and data.

Frequency Division Multiplexing
Frequency division multiplexing (FDM) divides the bandwidth of a transmission line into channels, where each channel transmits specific information. Figure 4.5 shows the multiplexing of several TV channels using FDM. The bandwidth of a coaxial cable is about 500 MHz and it can carry 80 TV channels. Each TV channel is assigned a different frequency, each using 6 MHz of bandwidth. Therefore, FDM combines several signals for transmission on a single transmission line.

Statistical Packet Multiplexing
Statistical packet multiplexing (SPM) dynamically allocates bandwidth to the active input channels, resulting in very efficient bandwidth utilization. In SPM, an

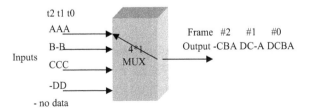

Fig. 4.4 TDM multiplexer

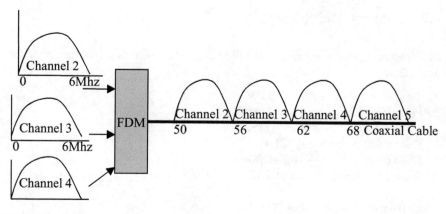

Fig. 4.5 Frequency division multiplexing (FDM)

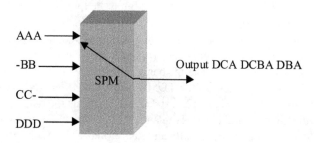

Fig. 4.6 Statistical packet multiplexer

idle channel does not receive any time allocation, as shown in Fig. 4.6. SPM uses a store-and-forward mechanism in order to detect and correct any error from incoming packets.

Fast Packet Multiplexing

Fast packet multiplexing (FPM) uses the same method as SPM and can assign maximum bandwidth to any input needed. FPM does not use a store and forward mechanism and, therefore, cannot perform error detection and correction. FPM will forward a packet before it has been completely received by the multiplexer.

Code Division Multiplexing (CDM)

In time division multiplexing (TDM), each end user is allocated a time slot for transmission. For instance, if 10 users are connected to a TDM and the bandwidth of the transmission link is 10 Mbps, then each user can transmit at the rate of only 1 Mbps. One disadvantage of TDM is that each user must wait for its turn to transmit its information. **Code division multiplexing (CDM)** is similar to TDM but allows all users to transmit simultaneously.

CDM Operation In CDM, each bit is divided into multiple bits that are called chip bits. This is done by multiplying logical 1 with a chip sequence, or by assigning chip bits to each node to represent logical 1. Table 4.1 shows the chip bits that have

Table 4.1 Chip bits for nodes A and B

End node	Chip bits for logical 1	Bipolar representation of chip bits
A	0101	−1 +1 −1 +1
B	1100	+1 +1 −1 −1

assigned to each node to represent logical 1. The complement of the chip bits represents logical zero. Chip bits can be represented by a bipolar value, so that +1 represents logical 1 and −1 represents logical zero.

Characteristics of Chip Bits In general, the chip bit sequence for A can be represented by $A = (A_4\, A_3\, A_2\, A_1)$ and the chip bit sequence for B can be represented by $B = (B_4\, B_3\, B_2\, B_1)$. One property of chip bits is that the inner product of two different chip bit sequences is zero and the inner product of the two identical chip sequences bits is one.

The inner product A and B is represented by and is defined by Eq. 4.1:

$$A \cdot B = \frac{1}{m}\sum_{i=1}^{4} A_i B_i = \frac{1}{4}\left(A_1 B_1 + A_2 B_2 + A_3 B_3 + A_4 B_4\right) \tag{4.1}$$

Therefore, $A \cdot B = \frac{1}{4}(-1+1-1+1)(+1+1-1-1) = \frac{1}{4}(-1+1+1-1) = 0$.

And the inner product A with itself $A \cdot A$ is:

$$A \cdot A = \frac{1}{4}(-1+1-1+1)(-1+1-1+1) = \frac{1}{4}(+1+1+1+1) = 1$$

CDM Architecture Fig. 4.7 shows the general architecture of CDM with three inputs and one output. The nodes A, B, and C are the inputs, and the chip bit sequences for each input are 4 bits each. The chip bits of each input are added, and that sum is then transmitted over the communication link. At the receiver side, the inner product of *the sum of the chip bits* and *the chip sequence of an input node* is used to determine the data bits for a specific node.

Example 4.1 Table 4.2 shows data to be transmitted by the nodes A, B, and C, and their chip sequences.

(a) Find the output of CDM.
(b) Find the data bit that is transmitted by node A at the receiver side.

The chip bit sequence for each data node is represented by Table 4.3.
The sum of the data is then transmitted to the receiver side. At the receiver side, the receiver uses the chip sequence of a specific node to recover the original data by using the inner product. In order to recover user A's data, the inner product of A's chip bit sequence and the sum of the chip bits is shown by Table 4.4, where +1 represents 1 and −1 represents 0.

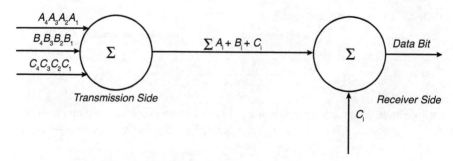

Fig. 4.7 General architecture of code division multiplexing

Table 4.2 Chip bits and data bits for nodes A, B, and C

Node	Chip bit sequence	Data to be transmitted
A	$-1 -1 -1 -1$	101
B	$-1 +1 -1 +1$	110
C	$+1 +1 -1 -1$	001

Table 4.3 Chip bits for Data of Node A, B, and C

Node A	$-1 -1 -1 -1 = (1)$	$+1 +1 +1 +1 = (0)$	$-1 -1 -1 -1 = (1)$
Node B	$-1 +1 -1 +1 = (1)$	$-1 +1 -1 +1 = (1)$	$+1 -1 +1 -1 = 0$
Node C	$-1 -1 +1 +1 = (0)$	$-1 -1 +1 +1 = (0)$	$+1 +1 -1 -1 = (1)$
Sum	$-3 -1 -1 +1$	$-1 +1 +1 +3$	$+1 -1 -1 -3$

Table 4.4 Inner product of the sum and node A's chip bits

Sum	$-3 -1 -1 +1$	$-1 +1 +1 +3$	$+1 -1 -1 -3$
Node A chip sequence	$-1 -1 -1 -1$	$-1 -1 -1 -1$	$-1 -1 -1 -1$
Inner product	$(+3-1 +1-1)/4 = +1$	$(+1 -1 -1 -3)$ $/4 = -1$	$(-1 +1 +1 +3)/4 = +1$

Wavelength Division Multiplexing

Wavelength division multiplexing (WDM) is used to transmit multiple rays of light with different wavelengths in one optical cable in order to increase the capacity of the optical cable, rather than using multiple optical cables. The concept of wave division multiplexing is similar to frequency division multiplexing. Figure 4.8 shows two optical rays with different wavelengths multiplexed and transmitted over an optical fiber cable.

Components of a WDM The components of a WDM are optical transponder, optical multiplexer, optical amplifier, and optical demultiplexer.

Optical Transponder The function of an optical transponder is to change the incoming ray's wavelength to another wavelength. Figure 4.9 shows the block diagram of an optical transponder.

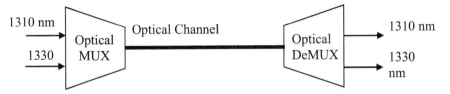

Fig. 4.8 Schematic diagram representing a simple form of WDM

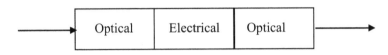

Fig. 4.9 Block diagram of an optical transponder

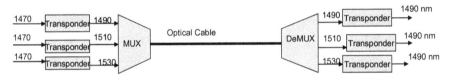

Fig. 4.10   Coarse division multiplexing

There are two types of WDM: dense wavelength division multiplexing (DWDM) and coarse wave division multiplexing (CWDM).

Dense Wavelength Division Multiplexing (DWDM) In DWDM, the wavelengths of the optical signals are close together. Current DWDM can transmit 60–80 wavelengths per channel with the wavelength spacing about 0.8 nm.

Coarse Wavelength Division Multiplexing (CWDM) CWDM can transmit 4–8 wavelengths of optical signals which are spaced 20 nm apart from each other. The ITU specifies 18 channels that use wavelengths from 1270 to 1610 nm for CWDM; however, some channels are not useable due to high signal attenuation. Figure 4.10 shows three optical rays with the same wavelengths connected to three optical transponders. The optical transponders change the wavelength of the incoming signals and are connected to the MUX. The outputs of the DMUX are connected to the three optical transponders in order to convert the signal wavelength back to their original values.

4.3   Telephone System Operation

The wired telephone system transmits information in analog form from a telephone set to the central office (CO). At the CO, the analog signal is converted to a digital signal, which is then transferred to the next central office as shown in Fig. 4.11. This digital signal is then converted to an analog signal and transmitted to the user. The method of conversion from analog to digital is called **pulse code modulation (PCM)**.

4.4   Digitizing Voice

Voice is an analog signal. In the central office of the telephone company, voices are digitized by a device called a **codec** (coder-decoder). The function of a codec is to digitize the voice signal and convert an already digitized signal to analog. According to the Nyquist theorem, in order to convert an analog signal into a digital signal, the analog signal must be sampled at least at the rate of two times its highest frequency. The voice signal must be sampled at 8000 samples per second because human speech is below 4000 Hz, as shown in Fig. 4.12. This method is called **pulse amplitude modulation (PAM)**.

Each PAM sample is represented by eight bits. In Fig. 4.13, it is represented by four bits. Remember, this method of converting voice to digital signal is called pulse code modulation (PCM). Since voice is digitized at the rate of 8000 samples per second and each sample is represented by 8 bits, the data rate of the human voice is 8000*8 = 64 kbps.

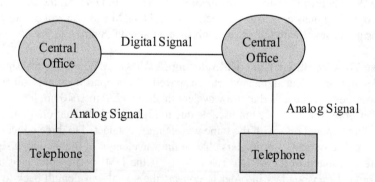

Fig. 4.11 Telephone system architecture

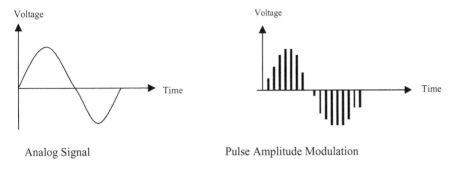

Fig. 4.12 Analog signal and pulse amplitude modulation (PAM). (a) Analog signal and (b) pulse amplitude modulation

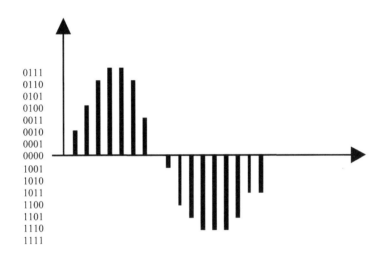

Fig. 4.13 Binary value for each PAM

4.5   T1 Links

Long-distance carriers use TDM to transmit voice signals over high-speed links. One of the applications of TDM is the T1 link. A **T1 link** carries a level-1 digital signal (DS-1). A DS-1 is generated by multiplexing 24 voice digital signals (digital signal level-0 or DS-0), as shown in Fig. 4.14. Pulse code modulation (PCM) is used to convert each analog signal to a digital signal. Each frame is made of 24 signals * 8 bits = 192 bits, with one extra bit added to separate each frame, making each frame 193 bits. Each frame represents 1/ 8000th of a second. Therefore, the data rate of T1 link is 193 * 8000 = 1.544 Mbps.

Table 4.5 shows TDM carrier standards for North America. Look at the table and you will see that a DS-2 can carry 96 voice channels with 168 Kbps overhead.

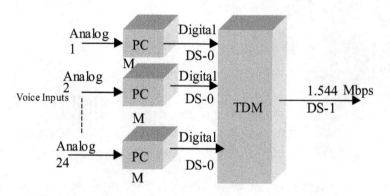

Fig. 4.14 Architecture of T1 Link

Table 4.5 TDM carrier standards for North America

Frame format	Line	Number of voice channels	Data rate (Mbps)
DS-1	T1	24	1.544
DS-1C	T-1C	48	3.152
DS-2	T2	96	6.312
DS-3	T3	672	44.736
DS-4	T4	4032	274.176

1 bit	Byte#24	Byte#23		Byte#2	Byte#1

Fig. 4.15 DS-1 frame format

Therefore, the data rate for DS-2 is 6.312 Mbps (96 channels * 64Kbps + 168 Kbps overhead). Figure 4.15 shows the DS-1 frame format where the 1-bit gap is used to separate each frame.

4.6   Switching Concepts

A communication network that has more than two computers must establish links between computers in order for them to be able to communicate with each other. One way to connect these computers is via fully connected network (mesh), as shown in Fig. 4.16.

The advantage of this method is that all stations can communicate with each other. The disadvantage is large number of connections are required when the number of stations is greater than four. To overcome this disadvantage, a device called a switch is used to connect stations, as shown in Fig. 4.17.

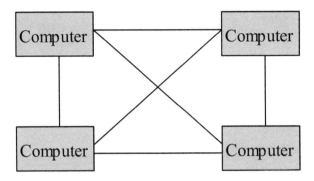

Fig. 4.16 Fully connected network

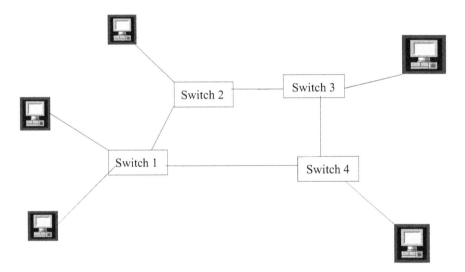

Fig. 4.17 Stations are connected by switches

The following types of switching are used in networking:

- Circuit switching
- Message switching
- Packet switching

Circuit Switching

In **circuit switching,** also called a connection-oriented circuit, a physical connection must be established for the duration of the transmission, such as in a telephone system. The application of circuit switching is for real-time communications. By dialing a telephone system, a connection is established, communication begins, and disconnection occurs at the end of the communication. Figure 4.18 shows circuit switching with multiple stations.

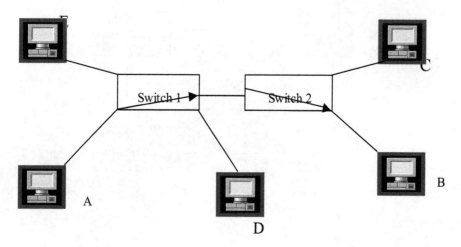

Fig. 4.18 Circuit switching

Advantages of Circuit Switching Circuit switching is used for real-time communication. There is no delay or congestion in the communication link because a physical connection exits between the source and the destination.

Disadvantage of Circuit Switching In circuit switching, only two stations can use the communication link at the same time. Therefore, it is not cost-effective. For example, in Fig. 4.18, if station A and B are communicating with each other, station D cannot communicate with station C. Station D must wait until A and B have finished their communication, then D may start communicating with C. In addition, if two stations such as A and B want to make a connection with C at the same time, a contention will occur and both stations must wait.

Message Switching
In message switching, station A sends its message to the switch. The switch stores that message and then forwards it to the destination. The disadvantage of message switching is that the switch needs to have a large buffer to store incoming messages from other links.

Packet Switching
Figure 4.19 shows a network with several switches. Assume source A has a message and wants to transfer it to destination B. Source A divides the message into packets and sends each packet, possibly by a different route. This process is known as **packet switching**. Each packet goes to the switch, which stores the packet and looks at the routing table inside the switch to find the next switch or destination. Each packet may take a different route and be received at the destination out of order. To prevent mistakes in reassembling the packets, each packet is given a sequence number which will be used by the destination to put the packets back in order.

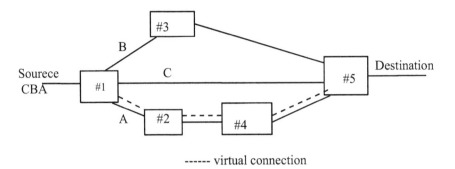

Fig. 4.19 Packet switching and virtual circuit

In Fig. 4.19, the source divides the message into 3 packets: A, B, and C. Then, the source transmits packet A to switch #1. Switch #1 stores packet A, looks at the congestion on all outgoing links, and finds that the link to switch #2 is the least congested. Switch #1 then sends packet A to switch #2. Switch #2 stores the packet and finds out from its routing table that packet A must go to Switch #4. Switch 2 then forwards the packet to switch# 4 and switch #4 forwards packet A to switch #5. Packet B and C take different routes, and as such, the packets might be received out of order. The destination uses the sequence numbers of the packets to put them in proper order. Packet switching is used to send information across the internet. This type of service is also called a connection-less oriented circuit.

Virtual Circuit
Virtual circuits are a type of packet switching which operate on the same concept as packet switching, but the routing of the packets is specified before transmission. As seen in Fig. 4.19, the source specifies the route, which is represented by dotted lines. Therefore, all the packets from source A go via the dotted line. By using this method, all packets will be received at the destination in the proper order.

Summary

- A multiplexer is used to share a transmission medium among several users.
- In time division multiplexing (TDM), the users are assigned equal time to use the digital channel.
- Frequency division multiplexing (FDM) is used for analog transmission, with the bandwidth of the analog channel divided into smaller channels.
- In statistical TDM, the bandwidth is dynamically allocated to active users.
- Code division multiplexing allows all user to transmit simultaneously.
- Wavelength division multiplexing is used for transmitting multiple ray of lights with different wavelengths over one optical cable.

- The pulse code modulation (PCM) method is used in the central switch to convert human voice to digital signal.
- The bandwidth of the human voice is 4000 Hz and it is digitized at a rate of 64 kbps.
- A T1 link is a special digital transmission line which has 24 inputs (each input is 64 kbps) and one output, with a data rate of 1.544 Mbps.
- There are three types of switching used in networking: circuit switching, packet switching, and virtual circuit.
- A message is divided into pieces. Each piece is called a packet.
- Packet switching treats each packet of a message separately.
- In circuit switching, a physical connection must be established between the source and destination before transmitting information.
- A virtual circuit is a type of packet switching. In a virtual circuit all packets of a message are transmitted in a specific path called a virtual path.

Key Terms

Circuit Switching	Packet Switching
Coarse Wave Division Multiplexing (CWDM)	Pulse Code Modulation (PCM)
Code Division Multiplexing (CDM)	Statistical Packet Multiplexing
Codec	T1 Link
Demultiplexer	Time Division Multiplexing (TDM)
Dense Wave Division Multiplexing (DWDM)	Transponder
Fast Packet Multiplexer (FPM)	Virtual Circuit
Frequency Division Multiplexing (FDM)	Wavelength Division Multiplexing (WDM)
Multiplexer	

Review Questions

Multiple Choice Questions

1. Several devices can share one transmission line by using a _____.

   (a) Multiplexer
   (b) Demultiplexer
   (c) BUS
   (d) CPU

2. _____ divides the bandwidth of a transmission line into channels.

   (a) TDM
   (b) FDM
   (c) SPM
   (d) FSPM

3. Code division multiplexing allows the users to transmit _____.

    (a) In an assigned time slot
    (b) One at the time
    (c) Simultaneously
    (d) None of the above

4. Wave division multiplexing is used for _____.

    (a) Optical signals
    (b) Analog signals
    (c) Digital signals
    (d) Radio frequency signals

5. The function of an optical transponder is to _____.

    (a) Change optical signals to electrical signals
    (b) Change the power of an optical signal
    (c) Change the wavelength of an optical signal
    (d) Change electrical signal to optical

6. _____ dynamically allocate bandwidth to active inputs.

    (a) TDM and FDM
    (b) SPM and FPM
    (c) FPM and TDM
    (d) FDM and SPM

7. What is the channel bandwidth of telephone system for the human voice?

    (a) 4000 Hz
    (b) 400 Hz
    (c) 40 Hz
    (d) 4 Hz

8. One of the applications of TDM is the _____ link.

    (a) Cable modem
    (b) T1
    (c) DSL modem
    (d) LAN

9. Virtual circuit is a type of _____.

    (a) Circuit switching
    (b) Packet switching
    (c) a and b
    (d) Message switching

10. Pulse code modulation is used to convert

    (a) Digital to analog
    (b) Analog to digital
    (c) Digital to digital
    (d) Analog to analog

11. Which of the following switching methods does deliver packets in order?

    (a) Packet switching
    (b) Virtual circuits
    (c) Circuit switching
    (d) b&c

12. The bandwidth of a telephone system is

    (a) 3 Khz
    (b) 4 KHz
    (c) 8 Khz
    (d) 40 Khz

Short Answer Questions

1. Show an 8-to-1 MUX and 1-to-8 DMUX.
2. List the types of multiplexers.
3. Explain the operation of TDM.
4. Describe the statistical packet multiplexer.
5. What is the difference between CDM and TDM?
6. What is an application of WDM?
7. What are the types of WDM?
8. What is the function of an optical transponder?
9. What is the type of signal used between two central switches of a telephone system?
10. What is the function of a Codec?
11. What does PCM stand for and what is its application?
12. Why must the human voice be sampled at the rate of 8000 samples per second?
13. How many voice channels does a T1 Link carry?
14. What type of multiplexer is used in a T1 link?
15. What is the data rate of the human voice and why?
16. What is the difference between a DS-1 and a T1 link?
17. What is the data rate of a T3 link?
18. How many voice channels can be carried by a T3 Link?
19. Explain the following switching operations:

    (a) Circuit switching
    (b) Message switching
    (c) Packet switching
    (d) Virtual circuit

20. Show the frame format of a T1 link.
21. Why is the data rate of a T1 link 1.54 Mbps?
22. The following inputs are connected to a 4*1 statistical multiplexer, show the outputs of the multiplexer:

    (a) Input #1 A – A – A
    (b) Input #2 B B B – B
    (c) Input #3 – C C – –
    (d) Input #4 D – D – D

23. What is the sampling rate of a signal with a highest frequency of 1000 Hz?
24. Use the chip sequences from Table 4.2 to find the data transmitted for Node C at the receiver side. Assume nodes A, B, and C have transmitted the following data:

Node A	111
Node B	010
Node C	001

Chapter 5
Error and Flow Control

Objectives

After completing this chapter, you should be able to:

- Comprehend frame transmission methods.
- Demonstrate understanding of error and flow control.
- Draw the Logical Link Control (LLC) frame format.

5.1 The Data Link Layer

As described in Chap. 1, the **Data Link layer** defines the frame format and type of transmission. The Data Link layer performs the following functions:

1. **On the transmitting side**: The Data Link layer accepts packets from the Network layer and breaks the information into frames. It then adds the destination MAC address, source MAC address, and Frame Check Sequence (FCS) fields, and passes each frame to the Physical layer for transmission.
2. **On the receiving side**: The Data Link layer accepts the bits from the Physical layer and forms them into a frame, performing error detection. If the frame is free of error, the Data Link layer passes the frame up to the Network layer.
3. **Frame synchronization**: This layer identifies the beginning and end of each frame.
4. **Flow control**: Distinguishes between control frames and information frames.
5. **Link management**: This layer coordinates transmission between the transmitter and receiver.
6. **Determine contention method**: This layer defines an access method in which two or more network devices compete for permission to transmit information across the same communication media, such as Token passing and Carrier Sense Multiple Access with Collision Detection (CSMA/CD).

A. Elahi, A. Cushman, *Computer Networks*,
https://doi.org/10.1007/978-3-031-42018-4_5

There are several existing protocols for the Data Link layer, such as:

- **Synchronous Data Link Control (SDLC)**: SDLC was developed by IBM as a link access for System Network Architecture (SNA).
- **High-Level Data Link Control (HDLC)**: HDLC is a version of SDLC modified by the ISO for use in the OSI model.
- **Link Access Procedure Balanced (LAPB)**: HDLC was modified by ITU and it is called LAPB used in ISDN.

5.2   Error and Flow Control

Functions of the Data Link layer include error detection, error control, and flow control. During the transmission of a frame from a source to its destination, the frame may get corrupted or lost. It is the function of the Data Link layer of the destination to check for error in the frame and inform the source about the status of the frame. This function must be performed in order for the source to retransmit the frame. One of the methods used is **Automatic Repeat Request (ARQ).** Positive or negative acknowledgement is used to establish reliable communication between the source and the destination. Automatic repeat request is carried out in two ways: Stop-and-Wait ARQ and Continuous ARQ.

Stop and Wait ARQ In Stop-and-Wait ARQ, the source transmits a frame and waits for a specific time for acknowledgement from the destination. If the source does not receive acknowledgment during this time, the source retransmits the frame. This method is used for networks with a half-duplex connection.

Case 1 The source station transmits a frame to the destination station. The destination station checks the frame for any errors. If there is no error in the frame, the destination station responds to the source station with a Positive Acknowledgment Frame ACK(N), where N is the sequence number of the frame. The source station transmits the next frame as shown in Fig. 5.1.

Case 2 The source station transmits a frame to the destination station. The destination station checks the frame for error. If there is an error in the frame, the destination station responds to the source station with a Negative Acknowledgment Frame NACK(N), where N is the sequence number of the corrupted frame. Then, the source retransmits the frame as shown in Fig. 5.2.

Case 3 The source station transmits a frame to the destination station. The source does not receive any acknowledgment due to the loss of the frame or loss of acknowledgment from the destination. When the source starts to transmit a frame to the destination, it sets a timer and waits for an acknowledgment. If the source does not receive an acknowledgement from the destination during that period of time, the frame is retransmitted, as shown in Fig. 5.3.

Fig. 5.1 Positive
acknowledgment

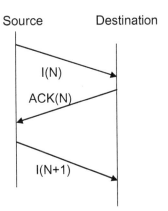

Fig. 5.2 Response to
corrupted frame by
destination

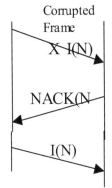

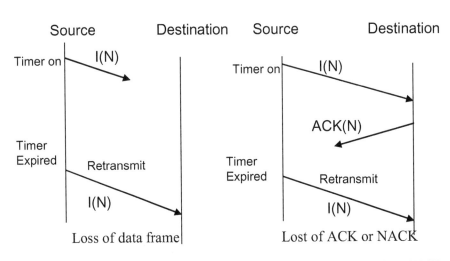

Fig. 5.3 Loss of ACK or NACK and I-frame. (**a**) Loss of data frame. (**b**) Loss of ACK or NACK

Continuous ARQ

In continuous ARQ, the transmitter continuously transmits frames to the destination. The destination sends ACK or NACK on different channels. Continuous ARQ is used in packet-switching network and full-duplex connection. There are two types of ARQ: Go-Back-N ARQ and Selective Reject ARQ.

Go-Back-N ARQ In the Go-Back-N ARQ method, the transmitter continuously transmits, and the receiver acknowledges each frame in a different channel as shown in Fig. 5.4.

Figure 5.4 shows that the source transmitted frame I5 and received NACK from I3. The source then retransmits frames I3, I4, and I5. In Go-Back-N, the source should hold a copy of those frames not receiving acknowledgment. When the source receives acknowledgment for a frame, it removes the frame from its buffer.

Selective Reject ARQ In selective ARQ, the source will retransmit only those frames for which the destination had sent a negative acknowledgment. Figure **5.5** shows the source transmitted frame I3 and received NACK1, which indicates that frame I1 was corrupted. The source retransmits only frame I1. In this method, the destination must have the capability to reorder frames that are out of order.

Sliding Window Method

In continuous ARQ, the source keeps a copy of transmitted frames in its buffer until it receives acknowledgment for a frame; it then removes the frame from its buffer. The continuous ARQ has the following deficiencies:

1. The destination may not have enough memory to store incoming frames.
2. The source may transmit frames faster than destination can process them.
3. The source must hold a copy of all unacknowledged transmitted frames in its buffer; therefore, the source requires a large buffer.
4. The file to be transmitted is divided into packets. Each packet has a sequence number; if the sequence number become large, it decreases network efficiency.

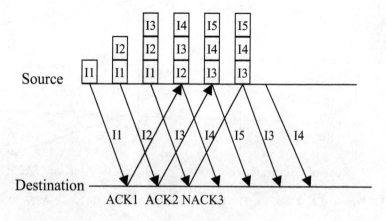

Fig. 5.4 Go-Back-N ARQ

Fig. 5.5 Selective
reject ARQ

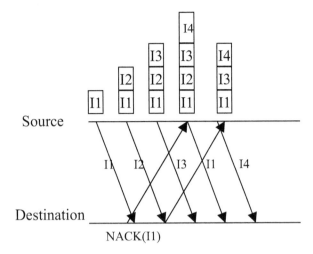

The Sliding Window Method limits the number of frames waiting for acknowl-edgement in source. For example, a source with a window of seven means the source can hold only seven unacknowledged frames in its buffer. The source will stop transmitting once it has seven frames in its buffer and wait for an acknowledgment frame. When the source receives acknowledgement for a frame, it removes that frame from its buffer and transmits the next frame. In order to prevent the need for large sequence numbers, most networking protocols use the following formula for assigning a sequence number to each frame:

$$\text{Sequence Number} = \text{Frame Number Modulo K}$$

(where K is the window size of the sending station).

For Example What is the sequence number for frame number 25? Assume that the window size is 7.

$$\text{Sequence Number} = 25 \text{ Modulo } 7 = 4$$

5.3   Frame Transmission Methods

Bit-Oriented Synchronization is one of the most popular frame transmission meth-ods, where information is transmitted bit by bit. Figure 5.6 shows the frame format of a bit synchronization frame. The start of the frame and the end of frame are rep-resented by eight bits in the form of 01111110. If the information field happens to contain the binary value 01111110, which normally indicates the end of a frame, an adjustment must be made by the transmitter. The transmitter inserts an extra zero

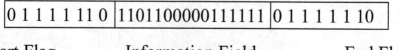

| 0 1 1 1 1 11 0 | 1101100000111111 | 0 1 1 1 1 1 10 |

Start Flag   Information Field   End Flag

Fig. 5.6 Format of bit-oriented synchronization

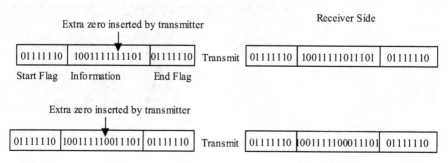

Fig. 5.7 Bit insertion in information field

any time five ones are repeated in the information field. The receiver will discard this extra zero. This technique is called bit insertion, as shown in Fig. 5.7.

5.4   IEEE 802 Standard Committee

The **IEEE 802 committee** defined standards for the Physical layer and the Data Link layer in February of 1980 and called it IEEE 802, with "80" representing 1980 and "2" representing the month of February. Figure 5.8 shows IEEE 802 standard and OSI model. The IEEE standard divides the data link layer of OSI model into two sub-layers: Logical Link Control (LLC) and Media Access Control (MAC).

Media Access Control (MAC)
The **Media Access Control** (MAC) layer defines the method that stations use to access the network, such as:

- Carrier Sense Multiple Access/Collision Detection (CSMA/CD) used for Ethernet
- Control Token used in Token Ring Networks and Token Bus Networks

Logical Link Control (LLC)
The **Logical Link Control (LLC)** defines the format of the frame. It is independent of network topology, transmission media, and Media Access Control. Figure 5.9 shows different MAC layers for several IEEE 802 networks. All networks which are listed use the same logical link control. Figure 5.10 shows the frame format of the LLC, which is used by all IEEE 802.X projects.

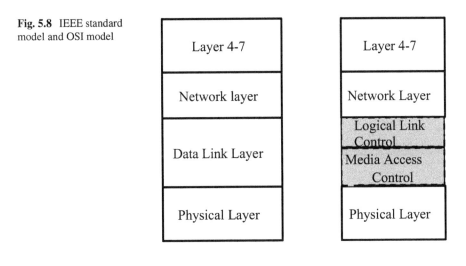

Fig. 5.8 IEEE standard model and OSI model

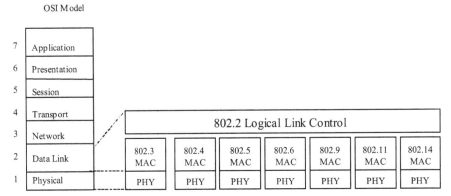

Fig. 5.9 IEEE 802 reference model

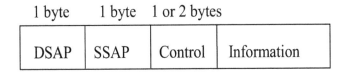

Fig. 5.10 Logical Link Control frame format

The following are the functions of each field of the LLC frame format:

Destination Service Access Point (DSAP) Since the destination station might run several network protocols such as Novell Netware, NetBIOS, Windows NT, and TCP/IP, the DSAP has to show the address of the protocol for the destination. Table 5.1 shows the most common value for service access point (SSAP and DSAP).

Table 5.1 DSSAP and SSAP Values

Protocol	SSAP and DSAP values In Hex
IBM SNA	04
IP	06
3Com	80
Novell	E0
Banyan	BC
Net BIOS	F8
LAN manager	F4

Source Service Access Point (SSAP) SSAP is a value of the source protocol and indicates the protocol that was used by the transmitter to send the packet.

Control Field The control field determines what type of information is stored in the information field, such as the information frame, supervisory frame, and unnumbered frame. The supervisory frames are receiver ready, receiver not ready, and reject. Some of the unnumbered frames are reset, frame reject, disconnect, and set asynchronous respond mode.

Summary

- The Data Link layer defines the frame format.
- The Network layer passes information to the Data Link layer, and then the Data Link layer adds the destination MAC address, source MAC address, and Frame Check Sequence (FCS).
- Frame synchronization is used to identify the beginning and end of each frame.
- Some Data Link protocols are SDLC, HDLC, and LAPB.
- A function of the Data Link layer is error detection and flow control.
- Types of flow control are Stop, Wait ARQ, and Continuous ARQ, as well as Sliding Window.
- Types of Continuous ARQ are Go-Back-N and Selective Reject ARQ.
- The Sliding Window method limits the number of frames waiting for acknowledgment.
- Bit-oriented synchronization uses bit-oriented transmission. Information is transmitted bit by bit.
- The IEEE 802 committee developed the standards for the Physical and Data Link layers.
- IEEE 802 divides the Data Link layer into two sub-layers: Logical Link Controls (LLC) and Media Access Control (MAC).
- All IEEE 802.X use the LLC frame format.

Key Terms

Automatic Repeat Request (ARQ)	Logical Link Control
Bit Oriented Synchronization	Selective Reject ARQ
Continuous ARQ	Sliding Window
Go-Back-N ARQ	Stop and Wait
IEEE 802 Committee	

Review Questions

Short Answer Questions

1. List the functions of the Data Link layer.
2. List three Data Link layer protocols.
3. List flow control methods.
4. What are the types of Continuous ARQ flow control?
5. IEEE 802 subdivides the Data Link layer into ___ sub-layers _____ and _____.
6. Show the frame format of IEEE 802.2.
7. Explain the function of the MAC layer.
8. What is a frame transmission method?
9. What does SDLC stand for?
10. What is an application of HDLC?
11. Explain bit insertion in bit-oriented synchronization.
12. Explain the function of synchronization bits.
13. Show the frame format of bit synchronization.
14. Explain Go-Back-N ARQ.
15. List three types of HDLC frame.
16. Show the transmitted frame after bit insertion for following frame:

$$0111111100000001111101111110$$

Chapter 6
Modulation Methods, Cable Modems, and FTTH

Objectives

After completing this chapter, you should be able to:

- Discuss modem operation.
- Explain the methods of signal modulation.
- Explain cable modem technology.
- Discuss fiber to the home (FTTH) operation.

6.1 Introduction

In order for two computers to communicate with each other, a link between them is required.

Currently, about 63 million households have cable TV services, and the same wire that brings TV signals to your house is a cable that can also provide Internet access with speed 100 times faster than a dial-up modem. The device that enables computers to access the Internet by cable TV lines is called a cable modem. The use of cable TV lines for this purpose is advantageous as digital signals cannot travel a long distance, but analog signals can.

6.2 Modem Operation

To link computers for communication over traditional telephone wires, a modem must be used, as shown in Fig. 6.1. A traditional telephone network operates with analog signals, whereas computers work with digital signals. Therefore a device is required to convert the computer's digital signal to an analog signal compatible with

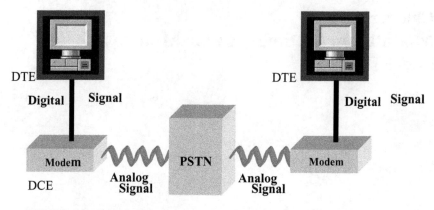

PSTN Public Switch Telephone

Fig. 6.1 Connection of two computers using a modem

the phone line (**modulation**). This device must also convert the incoming analog signal from the phone line to digital (**demodulation**). Such a device is called **a modem**.

A modem's transmission speed can be represented by either a data rate or baud rate. The **data rate** is the number of bits which a modem can transmit in 1 second. The **baud rate** is the number of signals which a modem can transmit in 1 second.

6.3   Modulation Methods

The carrier signal on a telephone line has a bandwidth of 4000 Hz. Figure 6.2 shows one cycle of a telephone carrier signal. The following types of modulation are used to convert digital signals to analog signals:

- Amplitude Shift Keying (ASK)
- Frequency Shift Keying (FSK)
- Phase Shift Keying (PSK)
- Quadrature Amplitude Modulation (QAM)

Amplitude Shift Keying (ASK)   In **Amplitude Shift Keying (ASK),** the amplitude of the signal changes. This is also referred to as Amplitude Modulation (AM). The receiver recognizes these modulation changes as voltage changes, as shown in Fig. 6.3. The smaller amplitude is represented by *zero* and the larger amplitude is represented by *one*. Each cycle is represented by one bit, with the maximum bits per second determined by the speed of the carrier signal. In this case, the baud rate is equal to the number of bits per second.

Fig. 6.2 Telephone carrier signal

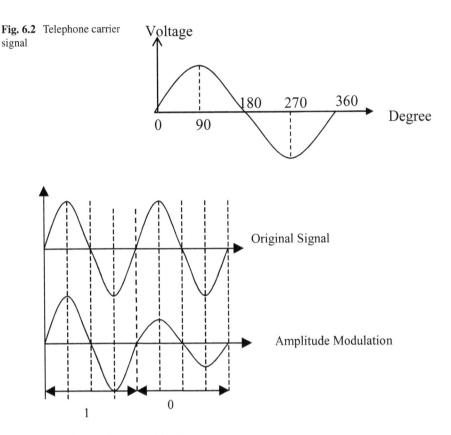

Fig. 6.3 Amplitude Shift Keying (ASK)

Frequency Shift Keying (FSK) In Frequency Shift Keying, a *zero* is represented by no change to the frequency of the original signal, while a *one* is represented by a change to the frequency of original signal. This is shown in Fig. 6.4. Frequency modulation is a term used in place of FSK.

Phase Shift Keying (PSK) Using the **Phase Shift Keying (PSK)** modulation method, the phase of the signal is changed to represent *ones* and *zeros*. Figure 6.5 shows a 90-degree phase shift. Figures 6.6a–c show the original signals with a 90-degree shift, a 180-degree shift and a 270-degree shift, respectively. Note that the original signal can be represented with four different signals: no shift, 90° shift, 180° shift, and 270° shift. Therefore, each signal can be represented by a two-bit binary number, as shown in Table 6.1.

The modem's speed using a 90-degree phase shift is 2*4000, which is equal to 8 Kbps. To increase the speed of the modem, the original signal can be shifted 45-degrees to generate eight distinct signals. Each signal can be represented by three bits. Therefore, the speed of the modem is increased to 3*4000, which is equal to 12 Kbps.

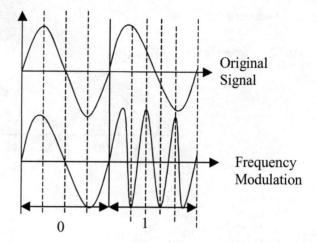

Fig. 6.4 Frequency Shift Keying (FSK)

Original Signal

Frequency Modulation

0   1

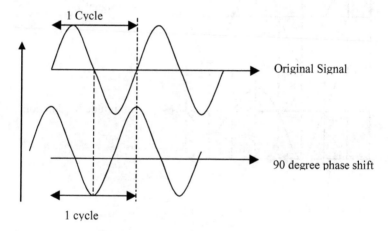

1 Cycle

Original Signal

90 degree phase shift

1 cycle

Fig. 6.5 90-degree phase shift

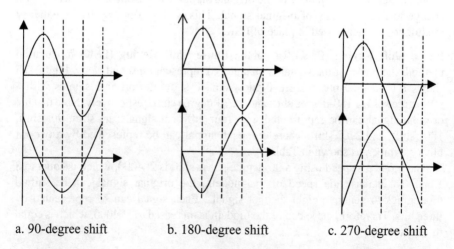

a. 90-degree shift   b. 180-degree shift   c. 270-degree shift

Fig. 6.6 Phase shift for 90, 180, and 270 degrees. (**a**) 90-degree shift, (**b**) 180-degree shift, and (**c**) 270-degree shift

Table 6.1 Phase shift and
binary value

Phase shift	Binary value
No shift	00
90°	01
180°	10
270°	11

Fig. 6.7 Constellation
diagram for Table 6.1

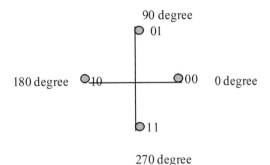

Fig. 6.8 Constellation diagram for 8-PSK

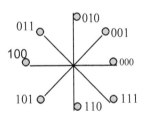

Bits	Phase Shift
000	0
001	45
010	90
011	135
100	180
101	225
110	270
111	315

The relation between phase and the binary representation of each phase can be plotted on a coordinate system called a **constellation diagram**. Figure 6.7 is a constellation diagram showing the four distinct signals of a 90-degree shift, with each signal represented by two bits. Figure 6.8 shows a constellation diagram using 45-degree shift and 3-bit representation (8-PSK).

Quadrature Amplitude Modulation (QAM) One method to increase the transmission speed of a modem is to combine PSK and ASK modulation. This hybrid modulation technique is called **Quadrature Amplitude Modulation (QAM)** and is shown in Fig. 6.9. Here we see the combination of four phases and two amplitudes which generates eight different signals called 8-QAM. Table 6.2 shows the binary value of each signal and provides a constellation diagram for 8-QAM. The data rate of this modem is 3 bits*4 K = 12 Kbps. Figure 6.10 shows the constellation diagram for such a modem.

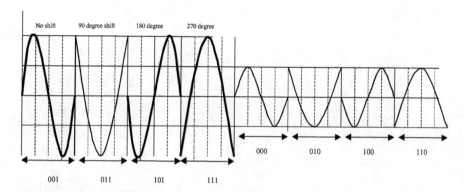

Fig. 6.9 8-QAM modulation

Table 6.2 Binary value
for 8QAM

Shift	Amplitude	Binary value
No	A1	000
No	A2	001
90°	A1	010
90°	A2	011
180°	A1	100
180°	A2	101
270°	A1	110
270°	A2	111

Fig. 6.10 Constellation
diagram for 8-QAM

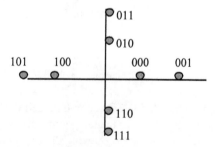

6.4   Cable Modem

The Cable Modem is another technology used for remote connection to the Internet. Residential access to the Internet is growing, and current modem technology can transfer data at only 56 kbps. Local telephone companies also offer a service known as Basic Rate ISDN, which has a transmission rate of 128 kbps. The cable modem offers high-speed access to the Internet using a media other than phone lines.

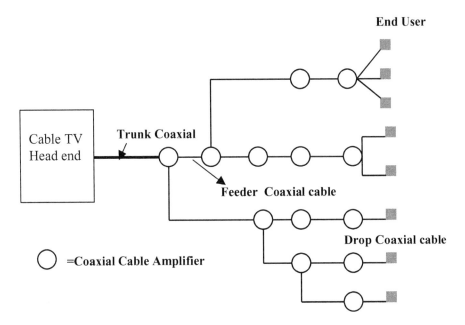

Fig. 6.11 Full coaxial cable TV system architecture

Cable TV System Architecture

Cable TV is designed to transmit broadband TV signals to homes using coaxial and fiber optic cable. Figure 6.11 shows the full coaxial cable TV system architecture. As shown in this diagram, cable TV uses Tree and Branch Bus Topology. The tree and branch cables are constructed of 75 ohms coaxial cable connected to the trunk cable.

The head end transmits TV signals over a **trunk cable** to a group of subscribers. The medium can be either coaxial or fiber-optic cable. The function of a **coaxial amplifier** is to amplify the signal, and it works in either direction. Feeder and drop cables are both coaxial cables. The **drop cable** is the part of a cable system that connects the subscriber to the feeder cable. **Feeder cables** are connected to trunk cable to cover a large area. The maximum distance between head end and subscriber is 10 to 15 km. The maximum number of cascaded amplifiers is 35 and the maximum number of connections is 125,000.

A TV signal transmits two frequency bands: VHF (very high frequency) and UHF (ultra-high frequency). The VHF channels use lower frequencies to generate stronger signals needed to transmit longer distances. The VHF channels start at channel 2 with frequency of 54 MHz and end at channel 13 with frequency of 216 MHz. UHF channels start at channel 14, with a frequency of 470 MHz and end at channel 83 with frequency of 890 MHZ. Each TV channel occupies 6 MHz of the TV radio frequency (RF) spectrum.

The bandwidth of coaxial cable is 500 MHz, with each TV channel requiring 6 MHz of bandwidth. The number of TV channels that can be transmitted is as

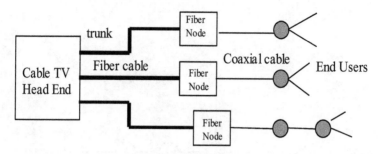

Fig. 6.12 HFC cable TV architecture

follows: (500–54)/6 = 75 channels. In order to increase the bandwidth of cable TV, cable TV Corporations use Hybrid Fiber Cable (HFC), which is a combination of fiber-optic cable and coaxial cable, as shown in Fig. 6.12. The bandwidth of a cable TV system using HFC cable is 750 MHz to 1GHz. Therefore, the number of channels that can be transmitted by a HFC cable can be computed by (750–54)/6 = 110 channels. The TV signal is transmitted to a fiber node using optical cable. The fiber node converts the optical signal to electrical signal and also converts electrical to optical. The coaxial amplifiers are two-way devices used to amplify the incoming signal. The maximum distance from head end to end user is 80 km and the maximum number of end user per fiber node connection is between 500 and 3000 (depending on the vendor). A channel between 5 MHz and 42 MHz is used to carry upstream signals from subscriber to the head end.

Cable Modem Technology

Figure 6.13 shows the components of a cable network consisting of a coaxial cable, **head end**, and a cable modem. The connection between cable modem and user is 10 Base-T. The user requires a 10Base-T NIC card to be able to use the cable modem. A cable modem can support more than one station using a router, as shown in Fig. 6.14.

A cable modem uses 64-QAM or 256-QAM modulation techniques to transmit information from the head end to the cable modem (downstream transmission). If a cable modem uses 256-QAM, this means 8 bits per signal and each signal is transmitted at 6 MHz. Theoretically the data rate of a cable modem is:

$$8^* \, 6Mhz = 48 \, Mbps$$

By using 64 QAM modulation, the data rate becomes

$$6^* 6Mhz = 36 \, Mbps$$

Upstream transmission (from cable modem to head end) uses a 2 MHz channel between 5 and 42 MHz. This low frequency is close to the CB radio frequency. The Quadrature Phase Shift Keying (QPSK) modulation method is used. The data rate of the cable modem for upstream transmission becomes:

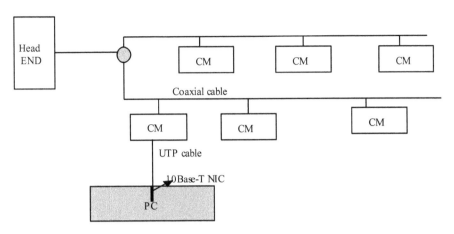

Fig. 6.13 Block diagram of a cable network

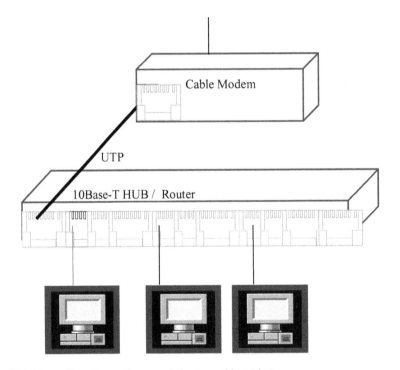

Fig. 6.14 Connection of more than one station to a cable modem

$$2^*600\text{khz} = 1200\,\text{kbps}.$$

Downstream and upstream bandwidths are shared by 500 to 5000 cable modem subscribers. If 100 subscribers are sharing a 36 Mbps connection, each user will receive a data at rate of 360Kbps. A cable modem provides a constant connection (like a LAN); it does not require any dialing. The cable modem head end communicates with the cable modem, and when the cable modem is commanded by cable modem head end, the modem will select an alternate channel for upstream transmission.

IEEE 802.14

A cable modem operates at the physical and data link layers of the OSI model. The **IEEE 802.14** standard provides a network logical reference model for the media access control (MAC) and physical layer. The following are general requirements defined for the cable modem by IEEE 802.14:

- Cable modems must support symmetrical and asymmetrical transmission in both directions.
- They support Operation, Administration, and Maintenance (OAM) functions.
- They support a maximum of 80 km distance for transmission from head end to the user.
- They support a large number of users.
- MAC layer should support multiple types of service, such as data, voice, and images.
- MAC layer must support unicast, multicast, and broadcast service.
- MAC layer should support fair arbitration for accessing the network.

6.5 Fiber to the Home (FTTH)

The demand for digital TV in the home, such as IPTV and HDTV, is growing fast. Most service providers offer bundle services that include HDTV, phone, and Internet services. This brings the question of how much bandwidth is needed to able to support standard digital TV, HDTV, phone, and Internet in a single home. Standard digital TV (SDTV) displays an image at a rate of 24 f/s (frames per second) and this rate is variable based on the country. In European countries, this rate is 25 f/s, while in North America it is 30 f/s. The SDTV frame is made of 650* 480 pixels, and to display a pixel, it requires two bytes. Therefore, the total bytes needed to display a frame are:

$$30 \, \text{frame} \, / \, \text{second}^* \, 650^* \, 450^* \, 16 \, \text{bits} = 140.4 \, \text{Mbps}$$

HDTV (high-definition TV) frames are made of 1920* 1080 pixels and use 3 bytes for displaying a pixel in color. Therefore, the required bandwidth for a single HDTV is:

$$30 \, \text{f} \, / \, \text{s}^* \, 1920^* \, 1080 \, \text{pixels}^* \, 24 = 1493 \, \text{Mbps}$$

The central head end (CO) compresses each frame before transmission. The most popular compression algorithms are MPEG-2, with typical compression ratio of 50/1, and MPEG-4, with compression ratio of 100/1.

SDTV uses MPEG- 2, so the bandwidth requirement becomes 140.4/50 = ~3 Mbps
HDTV uses MPEG-4, so the bandwidth requirement becomes 1493/100 = ~15 Mbps

A typical Internet user requires 2 Mbps.
The bandwidth retirement of VOIP is 100 kbps, which is negligible.
With this data, the bandwidth requirement for a house with two HDTVs, one SDTV, 4 Internet users, and a phone can be estimated as:

$$2^* 15 + 3 + 4^* 2 = 41 \text{Mbps}$$

In order for a service provider to provide 41 Mbps bandwidth, fiber to the home technology can be used.

FTTH Architecture
FTTH uses 100% fiber connections to the home and can be point to point (P2P) architectures (sometimes called an All-Optical Ethernet Network, or AOEN) or Passive Optical Networks. Figure 6.15 shows a P2P optical network, where a central office (CO) has direct connections to each house via fiber cable.

Passive Optical Networks
Passive optical networks (PON) use a single fiber connection from a central office, which is then split by a passive optical splitter as shown in Fig. 6.16. The function of splitter is to broadcast incoming rays to all of the outputs of the splitter. The maximum length of a feeder cable is 30,000 feet (9000 meters). If a splitter is a passive splitter, then it means it does not require any power. The maximum length of fiber cable from a splitter to the home is 3000 feet (900 meters). The splitter's input to output ratio can be 1/2, 1/4, 1/8, 1/16, or 1/32.

The splitter will reduce the power of the signal and that loss is a function of number of the outputs. The loss of the signal is represented in decibels, and the following equation can be used to determine the loss of signal in a splitter:

$$\text{Loss} = 3\text{db}^* \, N / 2, \text{where N is the number of outputs of the splitter}$$

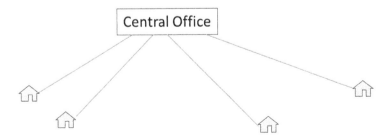

Fig. 6.15 Point to point (P2P) architecture

Fig. 6.16 Passive optical
network

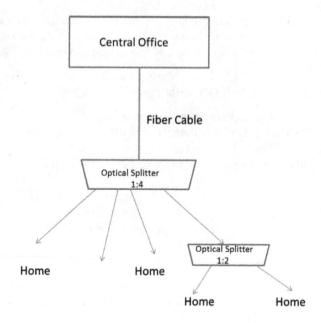

Table 6.3 Standards for FTTH

Technology	B-PON	G-PON	E-PON
Standard body	ITU G.983	ITU G.984	IEEE902.3ah
Data rate	155.52 Mbps upstream 155.52 or 622.08 downstream	2.44 Gbps upstream and down stream	1 Gbps upstream and down stream
Distance from OLT to ONU	20 km	10 to 20 km	10 and 20 km
Data format	ATM	ARM	Ethernet
Foreword error correction	Reed–Solomon error correction	Reed–Solomon error correction	None
Encryption	AES-128	AES-128	None

Most FTTH offers triple services such as voice, video, and Internet. The signals are transmitted upstream and downstream by using different wavelengths. There are three technologies that have been developed by standard bodies for FTTH (Table 6.3):

B-PON: Broadband Passive Optical Network
E-PON: Ethernet Passive optical Network
G-PON: Gigabit Passive Optical Network

Summary

- The function of a modem is to convert an analog signal to digital and digital signal to analog.
- Modulation methods are Amplitude Shift Keying (ASK), Frequency Shift Keying, (FSK), Phase Shift Keying (PSK), and Quadrature Amplitude Modulation (QAM).
- Amplitude Shift Keying (ASK) changes the amplitude of carrier signals in order to represent a digital signal.
- Frequency Shift Keying (FSK) changes the frequency of carrier signals in order to represent a digital signal.
- Phase Shift Keying (PSK) changes the phase of carrier signal.
- QAM modulation is a combination of ASK and PSK, used in high-speed modems.
- Baud rate is the number of signals per second that a modem can transmit.
- Data rate is the number of bits per second that a modem can transmit.
- Cable modems use a cable TV network to connect residential computers to the Internet.
- The head end of a cable TV system uses TV channels to transmit information to a cable modem at the subscriber site.
- Each cable TV channel requires 6 MHz bandwidth.
- Connecting a computer to a cable modem requires 10 Base-T network card.
- More than one station can be connected to a cable modem by using a hub or repeater.
- Cable modems operate in layer 1, layer 2, and layer 3 of the OSI model.
- IEEE 802.14 has developed the standard for cable modems.

Key Terms

Amplitude Shift Keying (ASK)	Hybrid Fiber Cable (HFC)
Baud Rate	IEEE 802.14 Standards
Coaxial Amplifier	Modem
Constellation Diagram	Modulation
Data Rate	Optical Splitter
Demodulation	Phase Shift Keying (PSK)
Downstream	Quadrature Amplitude Modulation (QAM)
Fiber to The Home (FTTH)	Trunk Cable
Head End	Upstream

Review Questions

Multiple Choice Questions

1. A modem converts_____.

   (a) Analog signal to digital
   (b) Digital signal to analog
   (c) a & b
   (d) Analog to analog

2. Cable TV is designed to transmit a _____ signal.

   (a) Baseband
   (b) Broadband
   (c) Digital
   (d) Optical signal

3. What is the data rate of a communication channel with bandwidth of 40 kHz and each signal is represented by 4 bits?

   (a) 40 Kbps
   (b) 80 Kbps
   (c) 160 Kbps
   (d) 10 Kbps

4. QAM modulation is a combination of:

   (a) ASK and FSK
   (b) ASK and PSK
   (c) PSK and FSK
   (d) None of the above

5. What type of modulation method is used in cable modems for downstream transmission?

   (a) DMT
   (b) QAM
   (c) QPSK
   (d) ASK

6. What type of modulation is used in cable modems for upstream transmission?

   (a) QAM
   (b) DMT
   (c) QPSK
   (d) FSK

7. What is the bandwidth of each TV channel?

    (a) 4 MHz

    (b) 2 MHz

    (c) 6 MHz

    (d) 1 MHz

8. What is the lowest frequency of TV Channel 2?

    (a) 40 MHz

    (b) 54 MHz

    (c) 60 MHz

    (d) 30 MHz

Short Answer Questions

1. What does modem stand for?
2. Explain the function of a modem.
3. Define baud rate.
4. Explain ASK modulation.
5. Explain FSK modulation.
6. Explain PSK modulation.
7. The speed of modem is represented in_____.
8. Distinguish between data rate and baud rate.
9. Draw a constellation diagram for 32QAM using 2 amplitudes.
10. What are the components of a cable TV system?
11. What does HFC stand for?
12. What is the bandwidth of a TV channel?
13. What type of modulation is used in cable TV modems for upstream transmission?
14. What is the type of modulation used in cable TV for downstream transmission?
15. What type of NIC is used in a computer connected to cable TV Modem?
16. What is the baud rate of ASK with data rate 600 bits per second?
17. What is the data rate of a modem using frequency shift keying with the baud rate of 300 signal per second?
18. What is data rate of a QAM signal with baud rate of 1200 and each signal represented by 4 bits?
19. Calculate the number of bits represented by each signal for a PSK signal with a data rate of 2400 bps and baud rate of 600.
20. How many bits per signal can be represented by a 32 QAM signal?
21. Calculate the baud rate of a 32 QAM signal with a data rate of 25 Kbps.
22. What does FTTH stand for?
23. List FTTH architectures.
24. What is the function of an optical splitter?

Chapter 7
Ethernet Technologies

Objectives

After completing this chapter, you should be able to:

- Describe Ethernet access methods.
- Discuss the function of each field in the Ethernet frame format.
- Distinguish between Unicast address, Multicast address, and Broadcast address.
- Explain the different types of Ethernet media.
- Discuss Fast Ethernet technology.
- Distinguish between the different types of Fast Ethernet media.
- Explain the differences and similarities between 100BaseT4, 100BaseTX, and 100BaseFX.
- Distinguish between different types of repeaters and know the maximum network diameter.
- Recognize Gigabit standards and the Gigabit Ethernet architecture.
- Identify the components of Gigabit Ethernet.
- Discuss the different types of gigabits Physical layers.
- List some of the applications for Gigabit Ethernet.
- List 10 GbE physical layers.
- Identify applications for 10 GbE.

7.1 Introduction

Ethernet was invented by the Xerox Corporation in 1972. It was further modified by Digital, Intel, and Xerox in 1980, which lead to Ethernet II or DIX (Digital, Intel, and Xerox). At that time, the IEEE (Institute of Electrical and Electronic Engineers) was assigned to develop a standard for Local Area Networks. The committee that standardized Ethernet, Token Ring, fiber optic, and other LAN technologies named

Fig. 7.1 Ethernet bus
topology

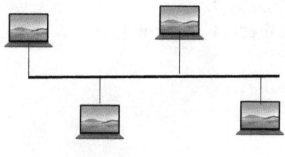

Fig. 7.2 Ethernet
reference model

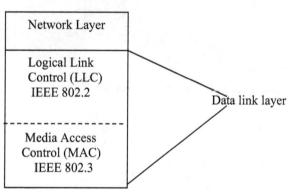

this family of LAN standards "802." The IEEE developed the standards for Ethernet
in 1984 and named them "IEEE 802.3." Ethernet uses the bus topology physically
and the star topology logically. It is still widely used as it is the least expensive LAN
to implement. Figure 7.1 shows an Ethernet Bus topology.

Figure 7.2 shows how Ethernet fits into the OSI model. The Data Link layer is
divided into two sublayers: the **Logical Link Control (LLC)** and the **Media Access
Control (MAC)** layers. The function of the LLC is to establish a logical connection
between source and destination. The IEEE standard for the LLC sublayer is IEEE
802.2. The function of the Media Access Control sublayer is to access the network,
which uses CSMA/CD (Carrier Sense and Multiple Access with Collision Detection).

7.2   Ethernet Operation

Each network card has a unique physical address. When a station transmits a frame
on the bus, all stations connected to the network will copy the frame. Each station
checks the address of the frame, and if it matches the station's NIC address, it will
accept the frame. Otherwise, the station discards the frame. In an Ethernet network,
each station uses the CSMA/CD protocol to access the network in order to transmit
information. CSMA/CD works as follows:

1. If a station wants to transmit, the station senses the channel (listens to the channel). If there is no carrier, the station transmits and checks for a collision as described in part 2. If the channel is in use, the station keeps listening until the channel becomes idle. When the channel becomes idle, the station starts transmitting again.
2. If two stations transmit frames at the same time on the bus, the frames will collide. The station which first detected the collision sends a jamming code on the bus (a jam signal is 32 bits of all ones), in order to inform the other stations that there is a collision on the bus.
3. The two stations which were involved in the collision wait according to a back-off algorithm (a method used to generate random waiting times for stations that were involved in a collision), and then start retransmission. Figure 7.3 shows the flowchart of CSMA/CD.

7.2.1   Ethernet II Frame Format

A block of data transmitted on the network, particularly with layer two traffic, is called a frame. There are two types of Ethernet frame formats: the Ethernet II frame format and the IEEE 802.3 frame format. These are shown in Fig. 7.4 and Fig. 7.6, respectively.

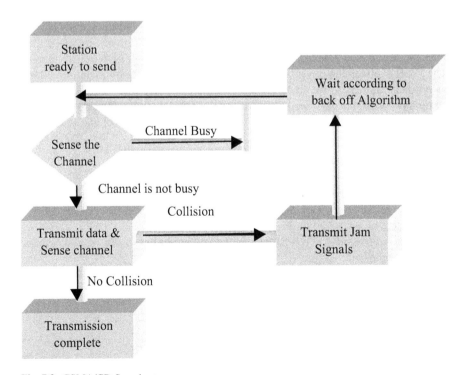

Fig. 7.3 CSMA/CD flowchart

Starting Delimiter (1 byte)	Destination Address (6 bytes)	Source Address (6 bytes)	Protocol Type (2 bytes)	Information field 46-1500	Frame Check Sequence (4 bytes)

Fig. 7.4 Ethernet II frame format

Fig. 7.5 Format of physical address

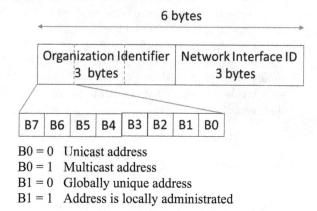

B0 = 0   Unicast address
B0 = 1   Multicast address
B1 = 0   Globally unique address
B1 = 1   Address is locally administrated

Start of Frame Delimiter (SFD) The SFD represents the start of a frame and is always set to 10101011.

Destination Address (DA) The destination address is the six-byte (48 bit) hardware address of a recipient station. This address is a unique address as no two are the same in the entire world. The hardware address of the Network Interface Card (NIC) is also called a MAC address (Media Access Control) or physical address. The IEEE oversees the physical addresses of NICs worldwide by assigning 22 bits of physical address to the manufacturers of Network Interface Cards. The 46-bit address is burned into the Read Only Memory (ROM) of each NIC and is called the universal administered address. Figure 7.5 shows the format of the destination address with the following types of addresses:

- **Unicast:** Recipient is an individual station.
- **Multicast:** Recipients are a group of stations.
- **Broadcast:** Recipients are all stations in the network. The 48-bit destination address is set to all ones, meaning that the DA address is FFFFFFFFFFFF in Hex, for a broadcast address.
- **Source Address (SA):** The SA shows the address of the source from which the frame originated.
- **Protocol Type:** The Protocol Type field defines the type of protocol generation information. The following are some of the protocol type numbers.

  - 0×0800 IP Internet Protocol (IPv4)
  - 0×0806 Address Resolution Protocol (ARP)
  - 0×8035 Reverse Address Resolution Protocol (RARP)

- 0 × 809B AppleTalk (Ethertalk)
- 0 × 80F3 Appletalk
- 0 × 8100 (identifies IEEE 802.1Q tag)
- 0 × 8137 Novell IPX (alt) 0 × 8138 Novell
- 0 × 86DD Internet Protocol, Version 6 (IPv6)
- 0 × 8847 MPLS unicast
- 0 × 8848 MPLS multicast
- 0 × 8863 PPPoE Discovery Stage
- 0 × 8864 PPPoE Session Stage

Data Field According to Fig. 7.4, the data field contains the actual information. The IEEE specifies that the minimum size of data field must be 46 bytes, and the maximum size is 1500 bytes. If the data field is less than 46 bytes, then the MAC layer will add as many bytes needed to reach a total of 46 bytes in the pad field..

Pad Field If the information in the data field is less than 46 bytes, extra information is added in the pad field to increase the size to 46 bytes.

Frame Check Sequence (FCS) The FCS is used for error detection to determine if any information was corrupted during transmission. IEEE uses CRC-32 for error detection.

Figure 7.6 shows the IEEE 802.3 frame format. Currently, the manufacturers producing NICs prefer to use Ethernet II due to that standard having less fields, which results in faster processing of the frames.

Length Field The two-byte field defines the number of bytes in the data field.

Control Field The control field determines the type of information in the information field, such as the supervisory frame, the unnumbered frame, and the information frame.

The Preamble, SFD, DA, SA, PAD, and FCS fields of the Ethernet II frame format are similar to those of IEEE 802.3.

Destination Service Access Point (DSAP) The MAC layer passes information to the LLC layer, which must then determine which protocol the incoming information belongs to, such as IP, NetWare, or DecNet.

Source Service Access Point (SSAP) The SSAP determines which protocol is sent to the destination protocol, such as IP or DecNet.

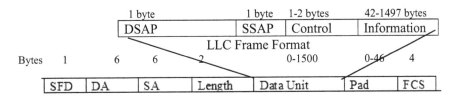

Fig. 7.6 IEEE 802.3 MAC and LLC frame formats

7.2.2   Ethernet Characteristics

The gap between each frame should not be less than 9.6 ms. A station can have a maximum of ten successive collisions. The size of the jam signal is 32 bits of all 1's. The maximum size of the frame is 1512 bytes including the header. Slot time is the propagation delay of the smallest frame. The smallest frame is 512 bits, and each bit time is 10^{-7} seconds; therefore, the propagation delay of the smallest possible frame is 512-bit time.

7.2.3   Ethernet Cabling and Components

The Ethernet network uses UTP media called 10BaseT, where the 10 defines the data rate, Base means Baseband transmission, and T means UTP cable. Figure 7.7 illustrates the port of a NIC which is used to connect a computer to a network.

10BaseT uses a UTP cable as transmission media and all stations are connected to a repeater or hub (switch), as shown in Fig. 7.8. The function of repeater (hub) is to accept frames from one port and retransmit the frames to all the other ports. Table 7.1 shows the pin connection of an RJ-45 connector.

The specifications of **10Base-T** are as follows:

- The maximum length of one segment is 100 m.
- The transceiver for 10BaseT is built into the NIC.
- Devices are connected to a 10BaseT hub in a physical star topology (while logically, they are in a Bus topology).
- A 10BaseT topology allows a maximum of four connected repeaters with a maximum diameter of 500 m.

Fig. 7.7 Network
Interface Card (NIC)

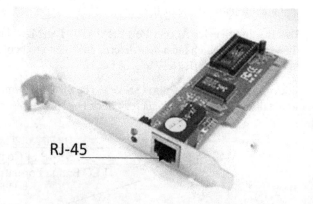

RJ-45

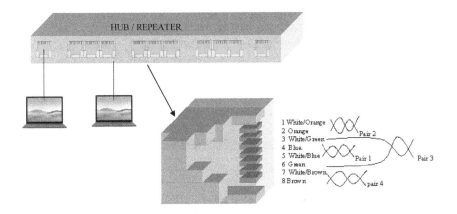

Fig. 7.8 10BaseT connections

Table 7.1 RJ-45
connector pins

PIN	Signals
1	RD+ pair 2
2	RD- pair 2
3	TD+ pair 3
4	NC pair1
5	NC pair1
6	TD- pair 3
7	NC pair 4
8	NC pair 4

7.2.4   UTP Cabling

There are two types of UTP cables used in networking. They are crossover and straight-through cables, as shown in Fig. 7.9.

A straight-through cable has identical ends and is used as a patch cord in Ethernet connections. A crossover cable is used to connect two Ethernet devices without a hub, or for connecting two hubs (Table 7.2).

7.3   Fast Ethernet Networking Technology

A group of leading network corporations formed a consortium to draft specifications for Fast Ethernet. This consortium proposed several of these specifications to the IEEE and **IEEE 802.3u** committee, which ultimately approved the standard for Fast Ethernet in 1995.

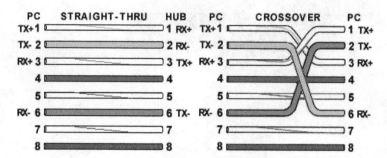

Fig. 7.9 Image of straight and crossover cables

Table 7.2 Application of straight-through and crossover cables

Device type	Device type	UTP cable type
Switch	PC	Straight through
Router	PC	Straight through
Switch	Switch	Crossover
PC	PC	Crossover
Router	Router	Crossover

Fast Ethernet is an extension of the Ethernet standard with a data rate of 100 Mbps, still using the Ethernet protocol. The goal of Fast Ethernet is to increase the bandwidth of Ethernet networks while using the same CSMA/CD transmission protocol. Using the same protocol for Fast Ethernet allows users to connect an existing 10BaseT LAN to a 100BaseT LAN with switching devices.

7.3.1 Fast Ethernet Media Types

One of the most popular media types for a Fast Ethernet network is unshielded twisted-pair wire, because it is easy to work with and it is a less expensive medium. The IEEE has approved specifications for the following three types of media for Fast Ethernet:

- 100BaseT4: 100 Mbps, Baseband, 4 pair Cat-3 cabling
- 100BaseTX: 100 Mbps, Baseband, Cat-5 cabling
- 100BaseFX: 100 Mbps, Baseband, fiber-optic cabling

100BaseTX technology supports 100 Mbps over two pairs of Cat-5 UTP cables. Cat-5 UTP cabling is the most common media for transmission and is designed to handle frequencies of up to 100 MHz. Manchester encoding, which is used for 10BaseT, is not suitable for 100BaseT because it doubles the frequency of the original signal. 100BaseT uses 4B/5B encoding with **Multiple Level Transition-3 (MLT-3)** levels for signal encoding. Figure 7.10 shows the hex value $(0E)_{16}$ converted from eight to ten bits (1111011100) using the 4B/5B encoding shown in

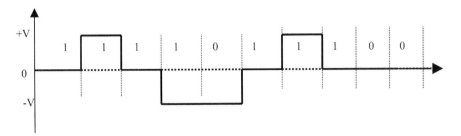

Fig. 7.10 MLT-3 signal for binary value 11110 11100

Table 7.3 4B/5B Encoding

4 bits binary	5 bits symbol		5 bits symbol
0000	11110	Idle	11111
0001	01001	Halt	00100
0010	10100	J	11000
0011	10101	K	10001
0100	01010	T	01101
0101	01011	Set	11001
0110	01110	Reset	00111
0111	01111	Quiet	00000
1000	10010		
1001	10011		
1010	10110		
1011	10111		
1100	11010		
1101	11011		
1110	11100		
1111	11101		

Table 7.3. That is then converted to MLT-3. MLT-3 reduces the frequency of the signal by a factor of four.

MLT encoding uses three voltage levels: +V, -V, and Zero. The MLT encoding rules are as follows:

1. If the next bit of the original signal is zero, then the next output is the same as the preceding value.
2. If the next bit of original signal is one, then the next output value has a transition (high to low or low to high).

   (a) If the preceding output was either +V or −V, then the next output value is zero.
   (b) If the preceding output was zero, then the next output is nonzero (it is the opposite sign of the last none-zero output).

100BaseFX technology transfers data at a rate of 100 Mbps using *fiber-optic* media for transmission. The standard cable for 100BaseFX is one pair of multimode fiber-optic cables with a 62.5-micron core and 125-micron cladding. The EIA

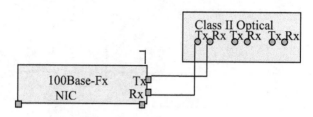

Fig. 7.11 100BaseFX connection to a repeater

recommends a SC plug-style connector. The SC connector uses push-on and push-off to connect and disconnect from the repeater. Figure 7.11 illustrates how 100BaseFX connects to a repeater.

100BaseFX uses 4B/5B encoding with NRZ-I signal encoding. In this type of encoding, four bits of information are converted to five bits, as shown again in Table 7.3, and the five bits are converted to NRZ-I digital signals. These signals are then converted to an optical ray for transmission over the fiber-optic cable to the receiver. At the receiver side, the optical signal is sampled every eight nanoseconds. If there is a change of light (from on to off or from off to on) in the sample, it is represented as a binary one. If there is no change of the light, it is represented by binary zero.

The conversion from four bits to five bits changes the data rate from 100 Mbps to 125 Mbps respectively. NRZ-I digital encoding reduces the frequency of transmission by half.

7.3.2   Fast Ethernet Repeaters

Repeaters are used to expand the network diameter. There are two types of repeaters used in Fast Ethernet: Class I repeaters and a Class II repeaters. The Class I repeater converts line signals from the incoming port to digital signals. This conversion allows different types of Fast Ethernet Technology to be connected to LAN segments. For example, it is possible to connect a 100BaseTX station to a 100BaseFX station by using a Class I Repeater. A Class II repeater repeats the incoming signal and sends it to every other port on the repeater. Most networks now use the switch, a layer two device, to expand a network.

7.4   Gigabit Ethernet Technology

With recent advances in the PCI bus and CPU technologies, workstations are getting faster. Today's PCI bus can transfer data at gigabit speed. A 64-bit PCI bus runs at 533 MHz and can transfer data at up to 6.4 gigabits per second. **Gigabit Ethernet** transfers data at one gigabit per second, or 10 times faster than Fast Ethernet. Gigabit Ethernet is technology compatible with Ethernet and Fast Ethernet, and it is used for backbones with gigabit switches.

Gigabit Ethernet is used for the campus **backbone** by connecting gigabit switches together. The switches operate in store-and-forward or cut-through technology. The IEEE 802 committee has developed a standard protocol called Quality of Service (IEEE 802.1p) which corresponds to the network layer of the OSI model. The IEEE802.1p standards provide tagging for each frame, indicating the priority or class of the service desired for the frame to be transmitted.

7.4.1   Gigabit Ethernet Standards

In 1995, the **IEEE 802.3** committee formed a study group called the IEEE 802.3z Task Force to research and develop standards for Gigabit Ethernet. In 1996, the Gigabit Ethernet Alliance was formed by more than 60 companies to support the development of Gigabit Ethernet.

7.4.2   Characteristics of Gigabit Ethernet

Gigabit Ethernet is used for linking Ethernet switches and Fast Ethernet switches, as well as for interconnecting very high-speed servers. Gigabit Ethernet enables organizations to upgrade their networks to 1000 Mbps while using the same operating systems and the same application software. The following are the characteristics of Gigabit Ethernet:

- Operates at 1000 Mbps (1 Gbps)
- Uses the IEEE 802.3 frame format and maximum frame size
- Supports full-duplex and half-duplex operation
- Uses the CSMA/CD access method for half-duplex operation and supports one repeater per collision domain
- Uses optical-fiber and copper wire for transmission media
- Supports 200-m collision domain diameters

7.4.3   Gigabit Ethernet Physical Layer

Figure 7.12 shows the Gigabit Ethernet physical layer, and Table 7.4 shows the different cable types and maximum distances for signal transmission over Gigabit Ethernet.

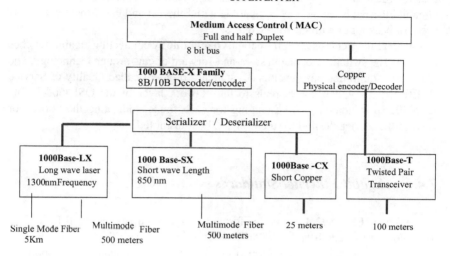

Fig. 7.12 Gigabit Ethernet physical layer

Table 7.4 Gigabit Ethernet cable types

Standard	Cable type	Core diameter in microns	Modal bandwidth MHz * km	Maximum distance in meters
1000BaseSX	MMF	62.5	160	220
	MMF	62.5	200	270
	MMF	50	400	500
	MMF	50	500	550
1000BaseLX	MMF	62.5	500	550
	MMF	50	400	550
	MMF	50	500	550
	SMF	9	N/A	5000
1000BaseCX	Twinax	–	–	25
1000BaseT	UTP	–	–	100

MMF means multimode fiber
SMF means single-mode fiber
SX means short wavelength of 850 nm
LX means long wavelength of 1300 nm

7.4.4   Gigabit MAC Layer

Gigabit Ethernet supports both half-duplex and full-duplex transmissions. Gigabit Ethernet with half-duplex uses the CSMA/CD access method while full-duplex uses a point-to-point connection. CSMA/CD defines the smallest frame for Ethernet as 64 bytes because the receiver should receive the first bit of the frame before the transmitter completes the transmission. By increasing the speed of the transmission from 100 Mbps to 1000 Mbps, the transmitter can complete the transmission before

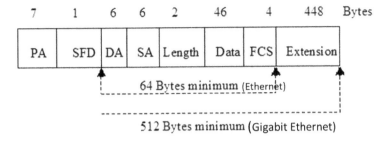

Fig. 7.13 Minimum frame size format for Gigabit Ethernet

the receiver receives the first bit of the frame. Fast Ethernet overcomes this problem by reducing the size of the cable, and Gigabit Ethernet increases the minimum size of the frame from 64 bytes to 512 bytes by adding carrier extensions to the Ethernet frame. Figure 7.13 shows the minimum frame size for Gigabit Ethernet.

7.5   10 Gigabit Ethernet

The IEEE 802.3ae task force completed the standard for **10 Gigabit Ethernet (10 GbE)** in March 2002. The 10 Gigabit Ethernet standard defines two types of physical layers: the **LAN physical layer (LAN PHY)** and the **WAN physical layer (WAN PHY)**. The WAN physical layer operates at the rate that is compatible with OC-192C and it uses Wave Division Multiplexing (WDM).

Applications of 10 Gigabit Ethernet for LANs are:

• Connecting a server to a switch with 10GbE
• Connections between switches

Applications of 10 Gigabit Ethernet for WAN are:

• Connecting two campus networks
• Storage Network Architecture (SNA)
• Connecting multiple networks in one metropolitan area with 10 GbE to offer services such as distance learning and video conferencing

7.5.1   Characteristics of 10 Gigabit Ethernet

The following points describe the characteristics of 10 Gigabit Ethernet:

• Uses the Ethernet frame format
• Uses the minimum and maximum Ethernet frame sizes
• Supports only full-duplex connections
• Uses optical cabling as a transmission medium

- Supports LAN and WAN physical layers
- Provides direct connection to OC-192C SONET

7.5.2 Gigabit Ethernet Physical Layer

Figure 7.14 shows the physical layer for 10 Gigabit Ethernet. It consists of serial transmission and Wave Division Multiplexing (WDM). Serial transmission uses different types of laser wavelengths. The following are different physical medium definitions for 10 Gigabit Ethernet. The 10 Gigabit Ethernet types and transmission distance are displayed in Table 7.5.

- 10GBASE- SR
- 10GBASE-SW
- 10GBASE- LR
- 10GBASE-LW
- 10GBASE-ER
- 10GBASE-EW
- 10GBASE-LX4

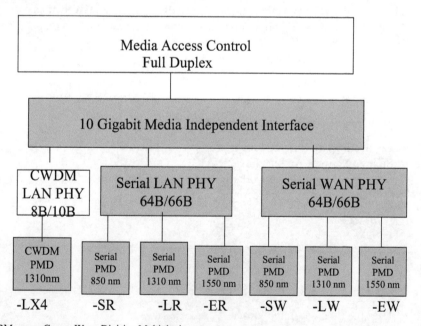

CWDM means Coarse Wave Division Multiplexing

Fig. 7.14 Physical layer of 10 Gigabit Ethernet. (CWDM means Coarse Wave Division Multiplexing)

Table 7.5 10GBE types and transmission distance

Standard	Cable types	Core diameter in microns	Model bandwidth	Distance in meters
10 GBASE-S	MM	50	500	66
10 GBASE-LX4	MM	62.5	160	300
10 GBASE-L	SM	9	–	10 km
10 GBASE-E	SM	9	–	40 km
10 GBASE-LX4	SM	9	–	10 km

The suffix for Gigabit Ethernet consists of three characters: the first character indicates the wavelength of the laser, S, L, or E, where:

S represents a short wavelength laser with a wavelength of 850 nm.
L represents a long wavelength laser with a wavelength of 1310 nm.
E represents an extended long wavelength laser with a wavelength of 1550 nm.

The second letter of the suffix represents the type of physical layer where:

R means physical layer for LAN.
W means physical layer for WAN.

Summary

- Ethernet and IEEE 802.3 use the bus topology logically.
- In the bus topology, the medium is shared by all stations.
- Ethernet uses Carrier Sense Multiple Access/Collision Detection (CSMA/CD) to access the network.
- Ethernet's data rate is 10 Mbps.
- The maximum size of an Ethernet frame is 1512 bytes.
- An Ethernet network card comes with three types of connectors: RJ-45, BNC, and DIX.
- 10BaseT is a medium with the following features: 10Mbps data rate, baseband, using UTP cabling with RJ-45 connectors.
- 10BaseT requires a repeater or hub.
- A Unicast address tells you that the recipient of the frame is an individual station.
- A Multicast address indicates that the recipient of the frame is a group of stations.
- A Broadcast address means the recipient of the frame is every station in the network.
- The data rate of Fast Ethernet is 100 Mbps.
- Fast Ethernet uses three types of media: 100BaseT4, 100BaseTX, and 100BaseFx.
- Fast Ethernet uses the same frame format as Ethernet.
- 100BaseT4 uses four pairs of Cat-3 UTP wires, 100BaseTX uses two pairs of Cat-5 UTP wires, and 100BaseFX uses fiber-optic cable.
- Fast Ethernet uses Class I repeaters to connect NIC cards with different types of media.

- Fast Ethernet uses Class II repeaters to connect stations having the same type of network Interface Card.
- Gigabit Ethernet has a data rate of 1000 Mbps.
- 10 Gigabit Ethernet has a data rate of 10,000 Mbps.

Key Terms

100BaseFx	Fast Ethernet Frame Format
100BaseT4	IEEE 802.3
100BaseTx	IEEE 802.3u
Bit Time	LLC Frame Format
Broadcast	Logical Link Control (LLC)
Class I Repeater	Media Access Control (MAC)
Class II Repeater	Multicast
CSMA/CD	Multilevel Transition (MLT-3)
Destination Address (DA)	Propagation Delay
Destination Service access Point (DSAP)	Repeater
Ethernet Frame Format	Source Address (SA)
Fast Ethernet	Source Service Access Point (SSAP)
Fast Ethernet Access Method	Unicast

Review Questions

Multiple Choice Questions

1. _____ based network is the least expensive LAN.

    (a) Ethernet
    (b) Token Ring
    (c) a and b
    (d) Gigabit Ethernet

2. The IEEE standard for Ethernet is _____.

    (a) IEEE 802.3
    (b) IEEE 802.4
    (c) IEEE 802.5
    (d) IEEE 802.2

3. Ethernet uses _____ to access the channel.

    (a) CSMA/CD
    (b) Token passing
    (c) Demand priority
    (d) Full duplex

4. A destination address in an Ethernet frame has _____bytes.

   (a) 2
   (b) 3
   (c) 6
   (d) 8

5. Ethernet uses _____ encoding.

   (a) Manchester
   (b) Differential Manchester
   (c) NRZ
   (d) NRZ-I

6. Fast Ethernet uses the _____ standard.

   (a) IEEE 802.2
   (b) IEEE 802.5
   (c) IEEE 802.3u
   (d) IEEE 802.4

7. The goal of Fast Ethernet is to increase _____.

   (a) The number of stations in a network
   (b) The frequency of signals
   (c) Bandwidth in a network
   (d) Network diameter

8. The role of the _____ is to interface the MAC sublayer to the physical medium dependent.

   (a) 100BaseT4
   (b) 100BaseTX
   (c) Convergence sub-layer
   (d) LLC

9. _____ is the most popular medium type for Fast Ethernet.

   (a) UTP
   (b) STP
   (c) Fiber optics
   (d) Coaxial cable

10. The data rate of 100BaseTX is _____ Mbps.

   (a) 100
   (b) 10
   (c) 200
   (d) 1000

11. 100BaseFX uses a _____ cable.

    (a) UTP
    (b) STP
    (c) Coaxial
    (d) Fiber-optic

12. There are _____ types of repeaters.

    (a) Five
    (b) Two
    (c) Three
    (d) Four

13. The maximum distance between two repeaters using a UTP cable is _____ meters.

    (a) 10
    (b) 5
    (c) 100
    (d) 200

14. Fast Ethernet's data rate is _____ Mbps.

    (a) 100
    (b) 10
    (c) 400
    (d) 200

15. What type of access method is used in Fast Ethernet?

    (a) Token
    (b) CSMA/CD
    (c) Demand priority
    (d) Full duplex

16. The data rate of Gigabit Ethernet is _____ Mbps.

    (a) 100
    (b) 1000
    (c) 200
    (d) 10,000.

17. The standard for Gigabit Ethernet is _____.

    (a) IEEE 802.2
    (b) IEEE 802.3
    (c) IEEE 802.3z (task force)
    (d) IEE802.3u

18. Gigabit Ethernet uses _____ access method for half-duplex operation.

    (a) CSMA/CD
    (b) Token passing
    (c) Demand priority
    (d) None of the above

19. Gigabit Ethernet uses _____ encoding.

    (a) Manchester
    (b) Differential Manchester
    (c) 8B/10B
    (d) 4B/5B

20. 1000BaseFX uses _____ cable for transmission of data.

    (a) UTP
    (b) Fiber-optic cable
    (c) Coaxial
    (d) STP

21. What type of protocol should be added to Gigabit Ethernet in order to carry voice and video information?

    (a) TCP
    (b) IP
    (c) 802.1p
    (d) RSVP

22. Gigabit Ethernet can operate in

    (a) Full duplex
    (b) Half duplex
    (c) a and b
    (d) None of the above

23. Gigabit Ethernet uses the CSMA/CD access method for_____.

    (a) Half duplex
    (b) Full duplex
    (c) a and b
    (d) None of the above

24. What is the transmission medium for 1000BaseT?

    (a) Cat-5 UTP
    (b) Cat 4 UTP
    (c) Coaxial cable
    (d) Fiber cable

25. What is the maximum length of cable used for 1000BaseT?

    (a) 50 m
    (b) 100 m
    (c) 200 m
    (d) 1000 m

26. What type of fiber cable is used for Gigabit Ethernet?

    (a) Multimode fiber
    (b) Single mode
    (c) Both a and b
    (d) None of the above

27. Gigabit Ethernet is used for:

    (a) WAN
    (b) Campus backbonc
    (c) MAN
    (d) Internet

28. The data rate of 10 GbE is _____ Mbps.

    (a) 100
    (b) 1000
    (c) 10,000
    (d) 100,000

29. Gigabit Ethernet operates in _____ mode(s).

    (a) Half duplex
    (b) Full duplex
    (c) a and b
    (d) CSMA/CD

Short Answer Questions

1. Define 10BaseT.
2. What do UTP and STP stand for?
3. What is a 10BaseT topology?
4. Show the Ethernet II frame format and the function of each field.
5. Explain the function of a repeater or a hub.
6. Show the IEEE 802.3 frame format and function of each field.
7. Describe the access method for Ethernet.
8. What does CSMA/CD stand for?
9. What is IEEE 802.2?
10. What is a MAC address?
11. Explain collisions in Ethernet?
12. What is a jam signal?

13. Explain broadcast addresses.
14. Describe unicast addresses.
15. What is the size a network interface card address?
16. What is the application of CRC (Cyclic Redundancy Check)?
17. What is the maximum size of a frame for IEEE 802.3?
18. How many bits of a network address represent the manufacturer ID?
19. How do computers distinguish one another on an Ethernet network?
20. What happens when two or more computers simultaneously transmit frames on an Ethernet network?
21. What is the function of the FCS field in the Ethernet frame format?
22. What is the function of the back-off algorithm in an Ethernet network?
23. What is the function of the length field in an Ethernet frame?
24. What is function of the protocol type field in the Ethernet II frame format?
25. List the IEEE sublayers of the data link layer.
26. What is function of the pad field in the IEEE 802.3 frame format?
28. Explain the following terms:

   (a) 100BaseT4
   (b) 100BaseTX
   (c) 100BaseFX

29. What is the cable type of 100BaseTX?
30. What is the difference between 100BaseTX and 100BaseT4?
31. What is the application of a Class I repeater?
32. What is the application of a Class II repeater?
33. What is the maximum network diameter using two Class II repeaters in a 100BaseT network?
34. Name the IEEE committee that developed the standard for Fast Ethernet.
35. Identify and explain the access method for Fast Ethernet.
36. What are the types of media used for Fast Ethernet?
37. What type of signal encoding is used for 100BaseT4?
38. What type of signal encoding is used for 100BaseFX?
39. Convert $(84)_{16}$ to 5 bit symbols using 4B/5B encoding, and then show the corresponding MLT digital signals.
40. What is the IEEE standard number for Gigabit Ethernet?
41. What is the data rate for Gigabit Ethernet?
42. What type of frame is used by Gigabit Ethernet?
43. What are the access methods for Gigabit Ethernet?
44. List the kinds of transmission media used for Gigabit Ethernet.
45. Explain the following terms:

   (a) 1000Base-CX
   (b) 1000Base-Lx
   (c) 1000Base-SX

46. What are the hardware components of Gigabit Ethernet?
47. What are some applications of 10 GbE?
48. Explain the following terms:

    (a) 10GBASE- SR
    (b) 10GBASE-SW
    (c) 10GBASE- LR
    (d) 10GBASE-LW
    (e) 10GBASE-ER
    (f) 10GBASE-EW

Chapter 8
LAN Interconnection Devices

Objectives

After completing this chapter, you should be able to:

- List LAN Interconnection devices.
- Describe the function and operation of a repeater.
- Describe the function and application of a bridge.
- Explain switch operation

    - Discuss the applications of LAN switching
    - Distinguish between symmetric and asymmetric switches

- Identify the application of a L2 switch, L3 switch and L4 switch.
- Discuss the application of virtual LANs.
- Understand the function of a router and the layers of the OSI model corresponding to a router.
- Describe the function and application of a gateway.

Introduction

Local area network (LAN) interconnection devices are used to expand the LAN to cover a larger geographical area and divide the traffic load by internetworking. By linking local area networks (LANs) to form a single network, such as separate LANs of different floors of a building or LANs in separate buildings, networks can be connected so that all computers in one site are linked. The devices discussed in this chapter are used for linking LANs together and can be distinguished by the OSI (Open system interconnection) layer at which they are operating.

© The Author(s), under exclusive license to Springer Nature Switzerland AG 2024   141
A. Elahi, A. Cushman, *Computer Networks*,
https://doi.org/10.1007/978-3-031-42018-4_8

8.1   Repeaters

A **repeater** is a device used to connect several segments of an LAN and to extend the allowable length of a network. A repeater accepts traffic from its input port and then retransmits the traffic at its output port. A hub is a multiple output repeater. A repeater works in the physical layer of the OSI Model. Figure 8.1 shows a repeater connecting 3 PCs.

8.2   Bridges

A **bridge** is used to connect the similar segments of a network together (homogeneous networks) and operates in the data link layer, as shown in Fig. 8.2. Bridges forward frames based on the destination addresses of the frames and detect transmission errors.

Fig. 8.1 Connecting 3 PCs to one repeater

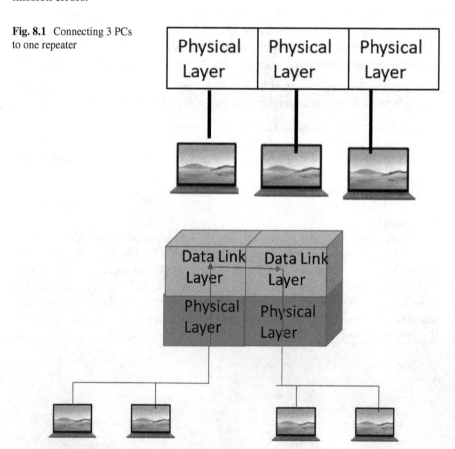

Fig. 8.2 OSI reference model of a bridge

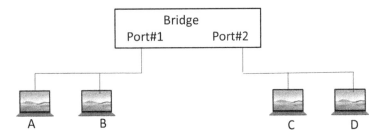

Fig. 8.3 Two Ethernet segments connected by a bridge

Functions of a Bridge
The function of a bridge is to analyze the incoming destination address of a frame and make a forwarding decision based on the location of the station. Figure 8.3 shows a bridge that is used to connect two Ethernet LANs together. For example, if station A sends a frame to station B, the bridge gets the frame and sees that station B is in the same segment as A and discards the frame. However, if station A forwards a frame to station C, the bridge would realize that station C is in a different LAN segment, so the bridge then forwards the frame to station C. The bridge forwards the data from one LAN to another without alteration of the frame. Bridges allow network administrators to segment their networks transparently, meaning that the individual station does not need to know that there is a bridge in the network.

Bridges are capable of *filtering*, which is useful for eliminating unnecessary broadcast frames. They can also be programmed not to forward frames from specific sources. By dividing a large network into segments and using a bridge to link the segments together, the throughput of the network will increase. If one segment of the network has failed, the other segments connected to the bridge can keep the network alive. Bridges also extend the length of the LAN. While stations A and B are communicating with each other, stations C and D can communicate with each other simultaneously.

Learning Bridge or Transparent Bridge   The learning bridge requires no initial programming. It can learn the location of each device by accepting a frame from the network segment and recording the MAC address and the port number. The frame comes to the bridge, which then retransmits the frame to all the segments of the network except the segment which sent the frame. By using this method, the learning bridge learns which station is connected to which segment of the network.

8.3   Switch

LAN switching is the fastest growing technology in the networking industry. Switches are used to connect LAN segments together to increase network throughput. A **switch** is a device with multiple ports which accepts packets from the ports of other computing devices. Switches can operate in layer 2 or layer 3.

Table 8.1 Switch
forwarding table

MAC address	Port number
MAC1	Port1
MAC2	Port2
MAC3	Port3
MAC4	Port4

Fig. 8.4 Switch with 4
ports

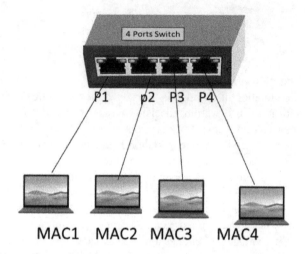

Layer 2 Switching (L2 Switch) A layer 2 switch (L2 switch) operates in the data link layer of the OSI model. It is used for network segmentation and for creating workgroups. The operation of a Layer 2 switch is similar to that of a multiport bridge, where a frame enters the switch from one port and is forwarded to the intended port based on the MAC address of the frame. A frame with a broadcast address will be repeated to all ports of the switch. During this process, a layer 2 switch learns the MAC addresses of the hosts connected to each port and creates a switching table which maps MAC addresses to port numbers, as seen in Table 8.1. The switch makes this table, called a switch forwarding table, through a learning method. When PC1 sends a packet to PC2, the switch recognizes that PC1 is connected to port 1, and when PC2 sends a packet to PC1 the switch recognizes that PC2 is connected to port 2. The switch then uses this table to forward frames to the proper ports. Figure 8.4 shows a 4-port switch with 4 PCs and their MAC addresses represented by MAC1 through MAC4.

Figure 8.5 shows two connected switches: switch 1 and switch 2. The forwarding table for switch 1 can be seen in Table 8.2. Assume that PC1 needs to send a packet to PC5. In order to forward the packet to PC5, switch 1 must check the destination MAC address of PC5 by consulting its switch forwarding table. Since the MAC address for PC5 is not listed in its table, switch 1 forwards the packet to default port 4, which then forwards the packet to default port 5 of switch 2. When the packet reaches switch 2, switch 2 then uses the information on its forwarding table to forward the packet to PC5.

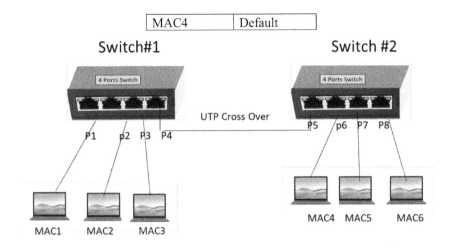

MAC4	Default

Fig. 8.5   Connecting two switches

Table 8.2   Switch forwarding
table for Fig. 8.5 switch #1

MAC address	Port number
MAC1	Port1
MAC2	Port2
MAC3	Port3
MAC4	Default

8.3.1   Spanning Tree Protocol (STP)

The spanning tree protocol is used to ensure a loop-free topology in a network with multiple switches or bridges. A switch will forward a broadcast and multicast frame to all of its ports. If there is a loop on the network, then the packet will travel in the loop continuously. Figure 8.6 shows a network with a connection loop.

In order to overcome loop problems, each switch runs the spanning tree algorithm (STP), also known as the standard IEEE 802.1d. The STP operation is described in the following steps:

1. Each switch is identified by an 8-byte ID. This ID is a combination of a two-byte priority field and the switch's 6-byte MAC address.
2. The switches exchange their IDs using Bridge Protocol Data Units (BPDU). The switch with the lowest ID is elected as the root bridge/switch.
3. The root bridge/switch places all its ports in the forwarding state.
4. Each nonroot bridge/switch finds the shortest path to the root bridge. The port used for shortest path to root bridge is call root port and is placed in forwarding state.
5. If a switch has more than one path to the root bridge, all ports not in the forwarding state are placed in the blocking state.

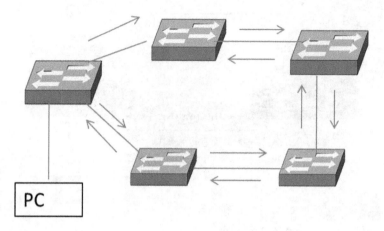

Fig. 8.6 Spanning tree operation

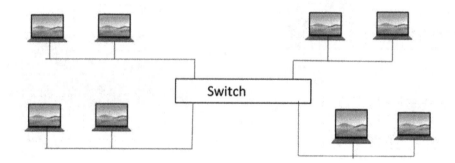

Fig. 8.7 Connection of LAN segments to a switch

8.3.2   Ethernet LAN Switching

Ethernet is one of the most popular LAN technologies because it uses unshielded twisted-pair cables. However, when the number of stations increases in an Ethernet LAN, the number of collisions also increases, and performance decreases accordingly. In order to increase the performance of an Ethernet LAN, it can be segmented with each of the segments connected to switch ports. In Fig. 8.7, each segment acts as an independent LAN and each segment similarly has its collision domain.

8.3.3   Switch Classifications

The manufacturers of switches classify the switches based on their applications: symmetrical and asymmetrical switching.

1. *Symmetric switching* provides switching between segments that have the same bandwidth. For example, 10Mbps to 10Mbps or 100Mbps to 100Mbps.
2. *Asymmetric switching* provides switching between segments of different bandwidths. For example, 10Mbps to 100Mbps or 100Mbps to 10Mbps.

8.3.4   Layer 3 Switch (L3 Switch)

A layer 3 switch (L3 switch) is a type of router that uses hardware rather than software. An L3 switch, sometimes called a routing switch, uses ASIC switching technology. This switch operates on the Network layer of the OSI model. The function of an L3 switch is to route the packet based on the logical address (layer 3) information. An L3 Switch accepts the packet from the incoming port and forwards the packet to the proper port based on a logical address, such as an IP address. In order to increase performance, the switch finds the route for the first packet and establishes a connection between the incoming and outgoing port for transferring the rest of the packets. This is called "route once and switch many."

8.4   Virtual LAN

A **virtual LAN (VLAN)**, also known as the IEEE 802.1q standard, is a configuration option on a LAN switch that allows network managers the flexibility to group or segment ports on an individual switch into logically defined LANs. There are two immediate benefits from a VLAN. First, it provides a way for network administrators to decrease the size of a broadcast domain and second, VLANs can provide security options for administrators. A VLAN is one way to prevent hosts on virtual segments from reaching one another. Another application of a VLAN is for logical segmentation of workgroups within an organization.

Port-Based VLAN In this method, VLAN membership is based on a switch port where the network administrator assigns each port of the switch to a specific VLAN ID. Only the stations connected to ports with same VLAN ID can communicate with each other. For example, Fig. 8.8 shows a four-port switch where ports 1 and 2 are assigned to VLAN ID 10 and ports 4 and 5 are assigned to VLAN ID 20. In this case, PC1 and PC2 can only communicate with each other and PC3 and PC4 can only communicate with each other.

8.4.1   VLAN Operation

A VLAN ID is assigned to each port of a switch. When a PC transmits a packet to a switch port, the switch inserts a tag into the packet. This tag includes the VLAN ID, as shown in Fig. 8.9. The switch then checks the VLAN ID of the packet and transmits it to the ports which have the same VLAN ID.

Fig. 8.8 A four-port
switch with 2 VLANs

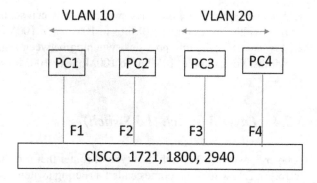

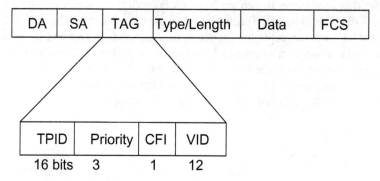

Fig. 8.9 IEEE 802.1q frame format

IEEE 802.1q developed a standard for the tagging of a frame for use with
VLAN. The IEEE 802.1q defines a method which allows a switch to add a tag to the
frame. It can process an untagged frame or a tagged frame. Figure 8.9 shows an
IEEE 802.1q frame format. The tag is 4 bytes and is inserted between the source
address (SA) and type/length field in the Ethernet frame format.

The functions of each subfield of the Tag field are described below:

TPID (Tag Protocol Identifier): This field is 16 bits and is set to 8100 (Hex) to
   identify the frame in IEEE 802.1q.
Priority: This field is 3 bits and identifies the priority of the frame.
CFI (Canonical Format Indicator): The CFI bit is mainly used for compatibility
   between Ethernet and token ring Networks and is set to 0 for Ethernet Switches.
VID (VLAN ID): This field is 12 bits and represents the VLAN number to which the
   frame belongs.

8.5   Routers

A **router** operates at the network layer of the OSI model to route a frame from one
LAN to another using a routing algorithm, as shown in Fig. 8.10.

Fig. 8.10 OSI reference
model for a router

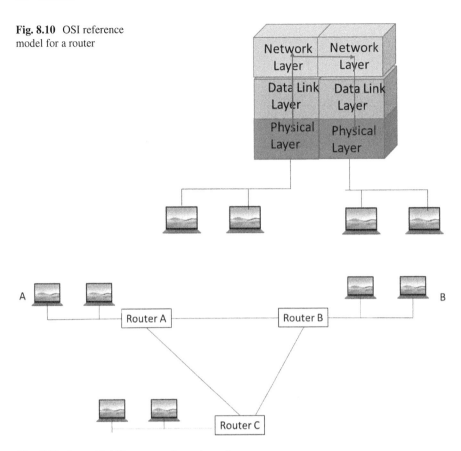

Fig. 8.11 Several LANs connected together using routers

The main function of a router is to determine the optimal data path and transfer information using that path. Figure 8.11 shows how routers can be used to connect several LANs together at different locations. Node A has a frame for Node B, so the Node A transmits the packet to Router A for it to find the best route to Router B, who then finally transmits the packet to Node B.

A router which can be configured manually by a network administrator is called a static router and a router that is configured by itself is called a dynamic router. In a static router, the routing table is administered manually by the network administrator who determines the route. In a dynamic router, the router uses routing algorithm to find the best route and updates it routing table automatically. The dynamic router also exchanges information with the next router on the network.

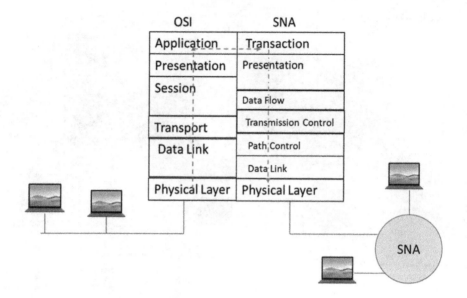

Fig. 8.12 Gateway connecting two different communication protocols

8.6   Gateways

Gateways operate up to the application layer, as shown in Fig. 8.12. The application of a **gateway** is to convert one protocol to another protocol. Figure 8.12 shows a network with IBM SNA (System Network Architecture) connected through a gateway, with an LAN running the TCP/IP protocol.

Summary

- LAN interconnection devices are repeaters, bridges, routers, switches, and gateways.
- A repeater is used to extend the length of the network and operates at the physical layer of an OSI model. A repeater accepts traffic from its input and repeats it at its output.
- A bridge is used to connect segments of same-type networks; the function of the bridge is to analyze the incoming frame's destination address and forward the frame to the proper segment. Bridges operate on the data link layer of the OSI model.
- A learning bridge or transparent bridge learns the location of each station by recording the NIC address and the port number of which frame enters the bridge.
- A source routing bridge routes the frame based on information in the routing field of the frame.

- A switch accepts a packet from one port and examines the destination address; it then retransmits the packet to the port having a host with the same destination address.
- When the number of users is increased in an Ethernet LAN, the number of collision will increase. To overcome this problem, Ethernet LAN can be segmented, with each segment connected to a port on a switch.
- Symmetric switch: It provides switching between LAN segments with the same data rate.
- Asymmetric switch: It provides switching between LAN segments with different data rates.
- Virtual LAN (VLAN): The IEEE802.10 committee approved the standard for VLAN. In VLAN, the switch port can be enabled and disabled by a network administrator. The administrator can also connect several ports to make a VLAN.
- Layer 2 switch: A multiport device that operates on layer 2 of the OSI model.
- Layer 3 switch: A type of router that uses integrated switching technology.
- A router is used to route a frame from one LAN to another LAN according to its routing table. Routers operate in the network layer of the OSI model.
- A gateway is used to convert one protocol to another protocol and operates in all seven layers of the OSI model.

Key Terms

Asymmetric switch	Proxy server
Bridge	Repeater
Dynamic router	Router
IEEE 802.1d	Source routing bridge
LAN interconnection	Static router
LAN switch	Switch
Layer 2 (L2) switch	Symmetric switch
Layer 3 (L3) switch	Transparent bridge
Learning bridge	Virtual LAN (VLAN)

Review Questions

Multiple Choice Questions

1. A hub is a multiple port _____.

   (a) server
   (b) switch
   (c) repeater
   (d) b & c

2. _____ operate in the data link layer.

   (a) Hubs
   (b) Repeaters
   (c) Switches
   (d) Gateways

3. _____ are capable of filtering.

   (a) Bridges
   (b) Repeaters
   (c) Switches
   (d) Hubs

4. In a _____, the frame contains the entire route to the destination.

   (a) source routing bridge
   (b) learning bridge
   (c) repeater
   (d) gateway

5. _____ are more complex internetworking devices than bridges.

   (a) Switches
   (b) Routers
   (c) Gateways
   (d) Hubs

6. A _____ operates up to the application layer.

   (a) router
   (b) switch
   (c) gateway
   (d) repeater

7. A _____ bridge learns the location of each station by recording the NIC address and the port number.

   (a) source routing
   (b) transparent
   (c) a and b
   (d) none of the above

8. A _____ is used to convert one protocol to another protocol.

   (a) router
   (b) switch
   (c) gateway
   (d) hub

9. The _____ is used to connect segments of a LAN.

   (a) router
   (b) hub
   (c) switch
   (d) gateway

10. A switch is a device with _____ port(s).

    (a) single
    (b) two
    (c) multiple
    (d) none of the above

11. A _____ provides switching between different bandwidth segments.

    (a) symmetric switch
    (b) asymmetric switch
    (c) gateway
    (d) router

12. Layer 3 switches or routing switches work on the OSI physical layer, data link layer, and _____ layer.

    (a) application
    (b) session
    (c) presentation
    (d) network

13. A _____ is a configuration option on a LAN switch.

    (a) VLAN
    (b) firewall
    (c) repeater
    (d) router

14. What type of switch is used to connect LAN segments within the same network? _____.

    (a) Layer 2 switch
    (b) Layer 3 switch
    (c) Layer 4 switch
    (d) All of the above

15. A layer 2 switch operates at the _____.

    (a) physical layer
    (b) data link layer
    (c) network layer
    (d) application layer

16. What type of switch is also used to route packets? _____.

    (a) Layer 2 switch
    (b) Layer 3 switch
    (c) Layer 4 switch
    (d) None of the above

Short Answer Questions

1. List LAN interconnection devices.
2. What is the function of a repeater?
3. Describe the function of a bridge?
4. What layer of OSI model does a bridge operate at?
5. Explain the operation of a transparent bridge.
6. Explain the operation of a source routing bridge.
7. Explain the function of a router.
8. Explain a static router.
9. What is the function of a router?
10. A router works in which layer of the OSI model?
11. Explain the definition of a dynamic router.
12. What is the application of a gateway?
13. A gateway operates in which layers of the OSI model?
14. What is the difference between a gateway and a router?
15. Explain the operation of a switch.
16. What is the application of a symmetric switch?
17. What is the application of an asymmetric switch?
18. What does VLAN stand for?
19. What is the difference between a router and a L3 Switch?
20. What is the function of spanning tree?

Chapter 9
Internet Protocols Part I

Objectives

After completion of this chapter, you should be able to:

- Discuss the history of the Internet.
- List the applications of the Internet and explain the function of each application protocol.
- Explain the function of the Internet Architecture Board (IAB).
- List Transmission Control Protocol and Internet Protocol (TCP/IP) and describe the service of each protocol.
- Distinguish between IP address classes and understand how IP addresses are assigned to a network of an organization.
- Show the TCP/IP reference model.
- Show the User Datagram Protocol (UDP) packet format and define the function of each field.
- List the applications protocol for Transmission Control Protocol (TCP).
- Describe the function of TCP, show the TCP packet format, and describe the function of each field.
- Explain the function of Internet Protocol (IP) and identify IP packet format.
- Explain TCP connection and disconnection.
- Show the IPv$_6$ format and explain the function of each field.
- Describe the advantages of IPv$_6$.

Introduction

The term Internet, short for Internetwork, describes a collection of networks that use the TCP/IP (Transmission Control Protocol and Internet Protocol) to communicate among nodes. These networks are connected through routers and gateways. Figure 9.1 shows an organization whose networks are connected by router to the Internet through an external gateway.

In 1968, the United States Department of Defense (DOD) created the Defense Advanced Research Project Agency (DARPA) for research on packet-switching

© The Author(s), under exclusive license to Springer Nature Switzerland AG 2024
A. Elahi, A. Cushman, *Computer Networks*,
https://doi.org/10.1007/978-3-031-42018-4_9

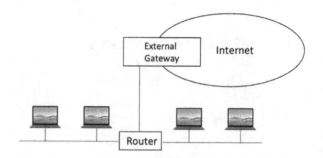

Fig. 9.1 Connection of a network to the Internet

networks. In 1969, DARPA created the **Advanced Research Project Agency (ARPA)**. In the same year, ARPA selected Bolt Beranek and Newman (BBN), a research firm in Cambridge, Massachusetts, to build an experimental network (ARPANET) to provide a test bed for emerging network technology. **ARPANET** originally connected four nodes, the University of California Los Angeles (UCLA), the University of California Berkeley (UCB), Stanford Research Institute (SRI), and the University of Utah, to share information and resources across long distances. ARPANET experienced rapid growth with the addition of universities. At that time, the protocol used in ARPANET was called the network control protocol (NCP). NCP did not scale well to the growing ARPANET, and in 1974, TCP/IP was introduced.

In 1980, the TCP/IP protocol became the only protocol that was in use on ARPANET. At the same time, most universities were using the UNIX operating system, which was created by Bell Labs in 1969. The University of California at Berkeley integrated the TCP/IP protocol into version 4.1 of its software distribution, later known as the Berkeley Software Distribution (Berkeley UNIX or BSD UNIX). The DOD then separated the military network (MILNET) from the nonmilitary network (ARPANET). In 1985, the National Science Foundation (NSF) connected the six supercomputer centers together and named this network **NSFNET**. The NSFNET was then connected to ARPANET. Naturally, NSF encouraged universities to connect to NSFNET. Due to the growth of NSFNET, in 1987, NSF accepted a joint proposal from IBM, MCI Corporation, and MERT Corporation to expand the NSF backbone. Figure 9.2 shows the NSF backbone in 1993. By 1995, numerous companies were running commercial networks.

The current Internet backbone consists of several backbones that belong to various Internet network service providers, such as MCI, AT&T, IBM, Sprint, and GTE. These backbones are connected through gateways. Figure 9.3 shows the GTE Internet backbone.

Internet Address Assignment
Any organization wishing to connect its network to the Internet must contact the Internet Network Information Center (InterNIC) to obtain an Internet Protocol address (IP address). The following are the Internet Network Information addresses:

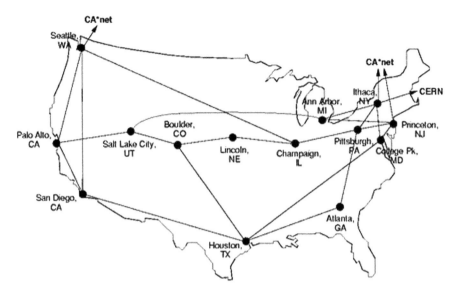

Fig. 9.2 NSFNET backbone

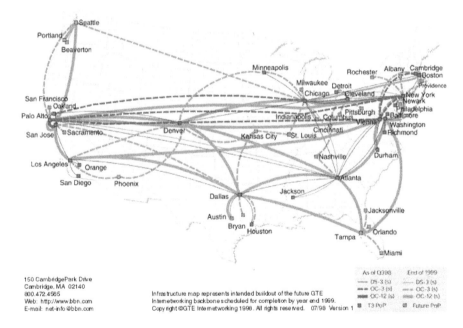

Fig. 9.3 GTE Internet backbone (http://www.bbn.com)

Website: www.internic.net
E-mail: Hostmaster@internic.net
Mailing address: Network Information Center
                     333 International
                     Menlo Park, CA 94025.

Any organization that obtains a network IP address will submit its server's name to InterNIC. InterNIC will ensure that no two servers have the same name. For Example: elahia1@ southernct.edu is the author's Internet address. Reading from right to left, the above domain name consists of:

edu: A top level domain which represents that Elahi is at some US educational site.
southernct: Represents the organization which owns the machine that has information about Elahi's IP address.

9.1   The Internet Architecture Board (IAB)

The Internet Architecture Board (IAB) is comprised of 13 members; six of whom are selected by the Internet Engineering Task Force (IETF). The functions of the IAB are the following:

- Determining the future of Internet addressing
- Architecture of the Internet
- Direction of IETF: Management of a top-level Domain Name System

The following is a list of the subcommittees of the IAB and their functions:

A. **IESG**: The Internet Engineering Steering Group works on Internet standards and oversees the work of all the other groups.
B. **IETF:** The Internet Engineering Task Force (IETF) is an open international committee of network designers, vendors and researchers. The IETF is divided into subgroups, which are organized based on their area of expertise, such as routing or transport.
C. **IRTF:** The function of the Internet Research Task Force (IRTF) is to promote long- and short-term research related to Internet protocols such as TCP/IP, Internet Architecture, and IPv6.
D. **IANA**: The Internet Assigned Numbers Authority (IANA) works under the Internet Network Information Center (INIC). The INIC consists of Network Solutions Inc. and AT&T Corp. The function of the INIC is registration of domain name servers (DNS) and education services. INIC manages registration of the second level domain names under the following top-level domains:

   gov, com, edu, net, mil, biz, info, name, coop, aero, net and **org.**

9.2   TCP/IP Reference Model

Transmission Control Protocol and Internet Protocol (TCP/IP) essentially consist of four levels: application level, Transport Level, Internet Level and Network Level, as shown in Fig. 9.4. Table 9.1 shows TCP/IP protocols and their functions.

9.3   TCP/IP Application Level

The application level enables the user to access the Internet. The following are common Internet applications:

- Simple Mail Transfer protocol (SMTP)
- Telnet
- File Transfer Protocol (FTP)
- Hyper Text Transfer Protocol (HTTP)
- Simple Network Management Protocol (SNMP)
- Domain Name System (DNS)

Simple Mail Transfer Protocol (SMTP)
SMTP is used for E-mail (electronic mail), which is used for transferring messages between two hosts. To send a mail message, the sender types in the address of the recipient and a message. The electronic mail application accepts the message/mail (if the address is right) and deposits it in the storage area/mailbox of the recipient. The recipient then retrieves the message from his/her mailbox.

An E-mail address (or just "email") is made up of a Username @ Mail Server Address. For example, in Elahia1@southernct.edu, "Elahia1" is the username and

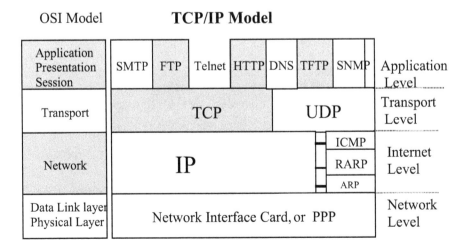

Fig. 9.4 TCP/IP reference model

Table 9.1 TCP/IP protocols and their functions

Protocol	Service
Internet Protocol IP	Provides packet delivery between networks
Internet Control Message Protocol ICMP	Controls transmission errors and control messages between hosts and gateways
Address Resolution Protocol ARP	Requests physical address from source
Reverse Address Resolution Protocol RARP	Response to the ARP
User Datagram Protocol UDP	Provides unreliable service between hosts (transfer data without acknowledgment)
Transmission Control Protocol TCP	Provides reliable service between hosts
Simple Network Management Protocol SNMP	Used for diagnostics purposes between hosts

"southernct.edu" is the domain name of the mail server. The "southernct" stands for Southern Connecticut State University, and "edu" stands for education.

Some email addresses are little more complicated, for example, Elahi@scsu1. southernct.edu. Here, "Elahi" is the username, "scsu1" is the name of a workstation that is a part of "southernct," and "edu" is the top-level domain representing an education center.

Telnet or Remote Login

Telnet is one of the most important Internet applications. It enables one computer to establish a connection to another computer. Users can login to a local computer and then remote login across the network to any other host. The computer establishing the connection is referred to as the local computer and the computer accepting the connection is referred to as the remote or host computer. The remote computer could be a hardwired terminal or a computer in another country. Once connected, the commands typed in by the user are executed on the remote computer. What the user sees on their monitor is what is taking place on the remote computer.

Remote login was originally developed for Berkeley UNIX to work with the UNIX operating system only, but it has since been ported to other operating systems. Telnet uses the client/server model. That is, a local computer uses a Telnet client program to establish the connection. The remote or host computer runs the Telnet server version to accept the connection and sends responses to requests.

File Transfer Protocol (FTP)

File Transfer Protocol (FTP) is an Internet standard for file transfer. It allows Internet users to transfer files from remote computers without having to log into them. FTP establishes a connection to a specified remote computer using FTP remote-host-address. Once connected, the remote host will ask the user for identification and a password. Upon compliance, the user can download or upload files.

Some sites make files available to the public. To access these files, users can enter *anonymous* or *guest* for identification and use his/her Internet address as a password. This application is called **anonymous FTP**.

Hyper Text Transfer Protocol (HTTP)
HTTP is an advanced file retrieving program that can access distributed and linked documents on the Web. Messages in HTTP are divided into request and response categories and work on the client/server principle. The request command is sent from the client to the server. The response command is sent from the server to client.

HTTP is a stateless protocol which treats each transaction independently. A connection is established between a client and a server for each transaction and is terminated as soon as the transaction is complete.

Simple Network Management Protocol (SNMP)
SNMP provides information for monitoring and controlling a network. It is used by network administrators to detect problems in networks such as issues with routers and gateways. SNMP is divided to the two parts; SNMP management system and SNMP agent. The SNMP management system issues commands to the SNMP agent, and the SNMP agent responds to the commend. The SNMP management system can mange network devices remotely.

DNS (Domain Name System)
DNS provides a way for users to easily remember websites. Memorizing a word such as "Yahoo" is much easier than memorizing numbers like "67.195.160.76"; therefore, the host name of a website is represented by a word rather than an IP address. However, for a client computer to access a website, send an E-mail, or connect to another computer through the Internet, the client computer must have the IP address of the destination. To obtain the IP address of the host, the client uses DNS to consult the DNS server and translate the word to the corresponding IP address.

9.4   Transport Level Protocols: UDP and TCP

The Transport level of the TCP/IP protocol consists of **UDP (User Datagram Protocol)** and **TCP (Transmission Control Protocol)**. The UDP protocol performs an unreliable connection service for receiving and transmitting data. TCP performs reliable delivery of data by adding a sequence number to each packet. When a packet reaches its destination, the destination acknowledges the sequence number of the next packet that it expects to receive.

9.4.1   User Datagram Protocol (UDP)

Some User Datagram Protocol (UDP) applications are Trivial File Transfer Protocol (TFTP) and Remote Call Procedure (RCP). UDP accepts information from the Application level and adds the source port, destination port, UDP length, and UDP

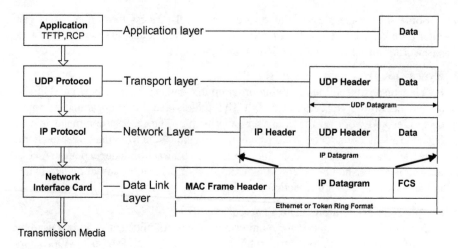

Fig. 9.5 UDP operation

0	31
Source Port 16 bits Define application, TFTP is port 69	**Destination Port** 16 bits Specifies Destination port on server
UDP Length 16 bits Define number of bytes in UDP header and data	**Checksum** 16 bits Checksum use for error detection of UDP header and data
DATA	

Fig. 9.6 UDP packet format

checksum. The resulting packet is called a UDP datagram packet with a total header length of eight bytes. The UDP protocol then passes the UDP packet to the IP. The IP adds its own header to the packet and passes the packet to the Logical Link Control (LLC). The LLC generates an 802.2 frame (LLC frame) and passes the LLC frame to the Medium Access Control (MAC) layer, which adds its own header and transfers the frame to the physical layer for transmission, as shown in Fig. 9.5.

UDP allows applications to exchange individual packets over a network as datagrams. A UDP packet sends information to the IP for delivery. There is no guaranteed reliability. Figure 9.6 shows the UDP packet format.

9.4.2 Transmission Control Protocol (TCP)

Most applications prefer to use reliable delivery of information. TCP offers reliable delivery of information through the Internet and gives users a way to transmit data in a reliable fashion. In TCP, before data are transmitted to the destination, a logical connection (not a physical connection) must be established before the information

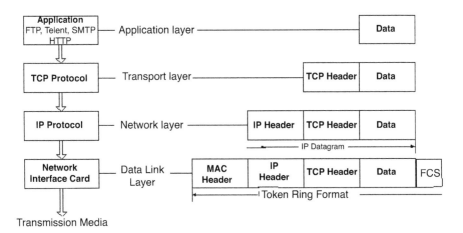

Fig. 9.7 TCP operation

is transmitted. TCP assigns a sequence number to each packet. The receiving end checks the sequence number of all packets to ensure that they are received. When the receiving end gets a packet, it responds to the destination by acknowledging the next sequence number. If the sending node does not receive an acknowledgment within a given time, it retransmits the previous packet.

Figure 9.7 shows application data passing through TCP. TCP adds a 20-byte header and passes it to the IP. The IP adds its own header and passes it to a Network Interface Card (NIC). The NIC adds a MAC header to the information and transmits the packet. Figure 9.8 shows the TCP packet format.

The following describes the function of each field in a TCP packet:

Sequence number: The number label for each packet sent by the source.

ACK sequence number: Acknowledges the next packet expected to be received from the source.

Header length: Identifies the length of the header in 32-bit word.

Flag bits: Six bits used for establishing a connection and disconnection.

URG   ACK   PSH   RST   SYN   FIN

URG: Urgent Pointer is set to "1" when that field contains urgent data.

ACK: ACK bit is set to "1" to represent that the acknowledge number is valid.

PSH: Set to "1" means the receiver should pass the data to an application as soon as possible.

RST: Resets connection.

SYN: Set to "1" when a node wants to establish a connection.

FIN: Set to "1" means this is the last packet.

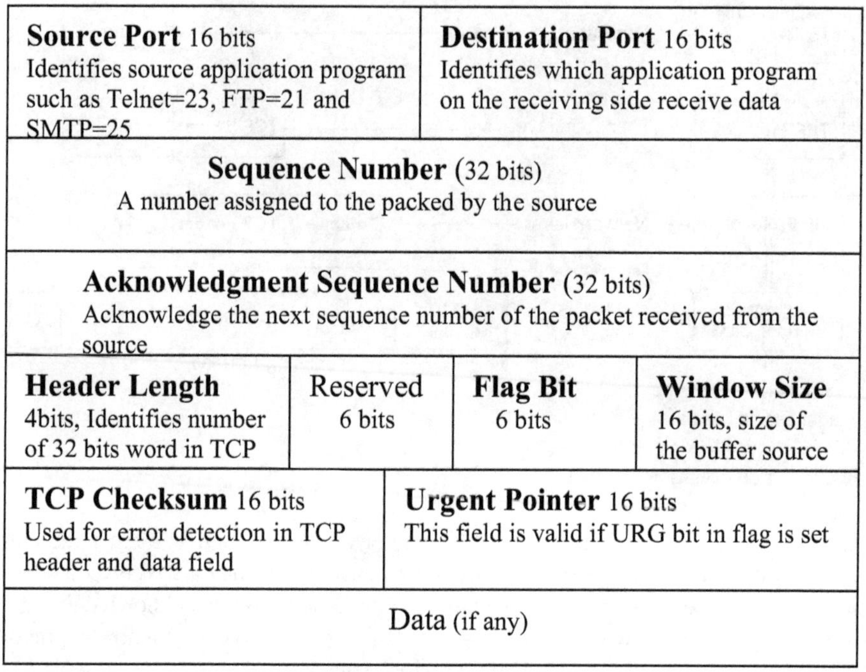

Source Port 16 bits Identifies source application program such as Telnet=23, FTP=21 and SMTP=25	**Destination Port** 16 bits Identifies which application program on the receiving side receive data		
Sequence Number (32 bits) A number assigned to the packed by the source			
Acknowledgment Sequence Number (32 bits) Acknowledge the next sequence number of the packet received from the source			
Header Length 4bits, Identifies number of 32 bits word in TCP	Reserved 6 bits	**Flag Bit** 6 bits	**Window Size** 16 bits, size of the buffer source
TCP Checksum 16 bits Used for error detection in TCP header and data field		**Urgent Pointer** 16 bits This field is valid if URG bit in flag is set	
Data (if any)			

Fig. 9.8 TCP packet format

Table 9.2 Commonly used port numbers

Network services	Port number/protocol	Network services	Port number/protocol
Tcpmux	1/tcp	netstat	15/tcp
Echo	7/tcp	ftp-data	20/tcp
Echo	7/udp	ftp	21/tcp
Discard	9/tcp	telnet	23/tcp
Discard	9/udp	smtp	25/tcp
Daytime	13/tcp	http	80/tcp
Daytime	13/udp		

Port Numbers

A **port number** is a logical channel in a communications system. The Transmission Control Protocol (TCP) and User Datagram Protocol (UDP) use port numbers to demultiplex messages to an application. Each application program has a unique port number associated with it. TCP/IP port numbers are between 1 and 65,535. The well-known ports are those from 0 to 1023 which are assigned by the IANA. Registered ports are those numbers from 1024 through 49,151. Dynamic or private port numbers range from 49,152 through 65,535. Table 9.2 shows a few commonly used port numbers.

9.5   Internet Level Protocols: IP and ARP

Internet level protocols consist of the IP, Address Resolution Protocol (ARP), Reverse ARP, and Internet Control Message Protocol (ICMP).

Internet Protocol Version 4(IPv4)

The function of IP is packet delivery with unreliable and connectionless service. These Internet datagrams are also called IP datagrams. All TCP, UDP, ICMP, and ARP data are transmitted as IP datagrams. Figure 9.9 shows an IP datagram packet format.

The following describes the function of each field in an IP packet:

Version: Contains 4-bit IP version number, and the current number, which is 4 (IPV_4).

Header length: Represents the number of 32-bit words in the header. If there are no IP options and padding, header length is 20 bytes (5 words). IP is an unreliable service so there is no acknowledgment from the destination to the source. There is also no physical connection between the source and destination. As such, IP datagrams can arrive at the destination out of order.

Type of services (TOS): TOS is 8 bits. For most purposes, the values of all bits in TOS are set to zero; meaning that the normal service unused bit is always zero.

Precedence indicates the importance of a datagram (0 is normal, 1 is next important). TCP/IP Protocol ignores this field.

31

IP Version 4 bits (current version is 4)	Header Length 4 bits Define number of 32-bit words in the header	Type of the Service(TOS) 8 bits specifies how the datagram should be handled		Total Length 16 bits specifies the length of IP datagram including the header in bytes.
Identification 16 bits used by destination to identify different datagram from one file	**Flags** 3 bits currently uses the first 2 bits DF and MF bits, DF=1 means do not fragment, MF=1 means More fragments are coming	**Fragment Offset** 13 bits contains the offset of the fragment from the beginning of the original datagram		
Time-to-Live TTL 8 bits specifies number of routers the datagram can pass	**Protocol** 8 bits specifies the protocol which data belongs to such as TCP, UDP, ICMP	**Header Checksum** 16 bits the 16 bit one's complement sum of the header		
Source IP Address 32 bits IP address of sending machine				
Destination IP address 32 bits IP address of receiving information				
Options if Any	**Padding**			
Data				

Fig. 9.9 IP datagram packet format

```
0      1       2 3 4 5 6        7
|  Precedence  |D|T|R| Unused |
```

D, T, and R identify the type of transport the datagram requests. D is for Delay, T is for throughput, and R is for reliability.

D = 0   Normal delay
D = 1   Low delay
T = 0   Normal throughput
T = 1   High throughput
R = 0   Normal reliability
R = 1   High reliability

Total length: This field identifies the total length of the datagram (including the header) in bytes.

Identification: This is a number created by the sending node. This number is required when reassembling fragmented messages. The identification field is used by the destination to put together related datagrams.

Fragmented offset field: The offset field represents the offset of data in multiples of eight; therefore, the fragment size should be multiples of eight.

Example: 1000 bytes are to be transferred over a network with an MTU of 256 bytes. Assume the header of each datagram is 20 bytes. Find the number of datagrams if the following information is given:

1. Identification: Can be any number
2. Total Length
3. Frame Offset
4. More Fragment

256–20 = 236 bytes
8 * 30 = 240
8 * 29 = 232. Each fragmented datum contains 232 bytes.

Identification	20	20	20	20	20
Total length Of each packet	232 + 20	232 + 20	232 + 20	232 + 20	72 + 20
Fragmented offset	0	29	58	87	116
MF	1	1	1	1	0

Example: 5480 bytes are being transferred from the FTP protocol to TCP for transmission. Show the fragment offset of the IP header:

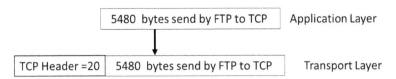

IP Protocol segments the Transport Layer to 1480   and add 20 bytes its header

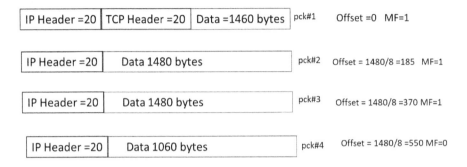

Time-to-Live (TTL): This field indicates the number of gateways or routers a packet can go through. This value is set by the sender (32 or 64) and is decremented by 1 every time a router handles the datagram. If this field becomes zero, the datagram is thrown away.

Protocol type: The number in this field identifies the High-Level Protocol that generates this datagram or allows the destination IP to pass the datagram on to the required protocol. The following are common protocol numbers (Table 9.3).

Header checksum: This field is the checksum of the header (not the data field). The checksum is the sum of the one's complement of the 16-bit word of the header.

Sending address: This is the IP address of the source.

Destination address: This is the IP address of the destination.

Maximum Transfer Unit (MTU)
The Maximum Transfer Unit (MTU) is the largest frame length that can possibly be sent over a given physical medium as there is a limit on frame size. For example, 802.3's maximum frame size is 1500 bytes. If the datagram is larger than the MTU, the datagram is fragmented into several frames, each less than the MTU. Table 9.4 shows MTU values for a few common network types:

Table 9.3 Protocols and their numbers

Protocol name	Protocol number
Internet Control Message Protocol (ICMP)	1
Transmission Control Protocol (TCP)	6
User Datagram Protocol (UDP)	17
General Routing Encapsulation (PPTP data over GRE)	47
Authentication Header (AH) IPSec	51
Encapsulation Security Payload (ESP) IPSec	50
Exterior Gateway Protocol (EGP)	8
Gateway-Gateway Protocol (GGP)	3
Host Monitoring Protocol (HMP)	20
Internet Group Management Protocol (IGMP)	88
MIT Remote Virtual Disk (RVD)	66
OSPF Open Shortest Path First	89
PARC Universal Packet Protocol (PUP)	12
Reliable Datagram Protocol (RDP)	27
Reservation Protocol (RSVP) QoS	46

Table 9.4 Network types with MTU values

Network type	MTU (bytes)
4 mbps token ring	4464
16 Mpbs token ring	17,914
FDDI	4352
Ethernet	1500
X.25	576
Point-to-point	296

9.6   IPv4 Addressing

An IPv4 address is a 32-bit number which forms a unique address for each host connected to the Internet. No two hosts can have the same IP address. The assignment and maintenance of IP addressing is maintained by InterNIC. An IP address is written in dotted decimal (Base$_{10}$) notation and is represented by four 8-bit binary numbers with the range of 0 to 255 ($4 \times 8 = 32$ bits).

Binary (Base$_2$)   00000000 to 11111111
Decimal (Base$_{10}$)   0   to   255

IP addresses are organized into the following five classes:

1. Class A IP Address

The Class A IP address, as seen in Fig. 9.10, is used for organizations with a large number of users connected to the Internet and a small number of networks.

7 bits	24 bits

0	NET ID	HOST ID

Fig. 9.10 Class A IP address format

The first most significant bit of a Class A IP address is zero.
The Network ID is 7 bits.
The Host ID is 24 bits.
The range of a Network ID is from 0 to 127. The numbers 0 and 127 are reserved.
The range of Class A addresses is from 0.0.0.0 to 127.255.255.255 in dotted decimal.

2. Class B IP Address

A Class B address is used for medium-sized networks having more than 255 hosts, as shown in Fig. 9.11.

The first two most significant bits of a Class B address are 1 and 0.
The Network ID is 14 bits.
The Host ID is 16 bits.
With a Class B address we can have 2^{14} (16,384) networks, and each network can have 2^{16} (65,536) hosts or nodes.
The range of a Class B Network ID is from 128.0. to 191.255.

3. Class C IP Address

A Class C address, as seen in Fig. 9.12 is used for networks with a small number of hosts (those networks whose number of hosts does not exceed 255).
The first 3 bits of a Class C address are 1, 1, and 0.
Twenty-one bits are used for Network ID, and 8 bits are used for the Host ID.
A Class C IP address can handle 2^{21} networks, where each network can have 256 host IDs.
The range of a Class C Network ID is from: The IP address of **192.0.2.1** was never assigned and used for *test* purposes only.

4. Class D IP Address

Class D address is reserved for multicasting. In multicasting, a packet is sent to a group of hosts.
The range of a Class D Network ID is from 224.0.0.0 to 239.255.255.255. The IP address format for Class D addresses can be seen in Fig. 9.13.

5. Class E IP Address

Reserved for research, the range of a Class E addresses is from 240.0.0.0 to 247.255.255.255.

14 bits		16 bits
1	0	NET ID	HOST ID

Fig. 9.11 Class B IP address format

21 bits		8 bits
1	1	0	NET ID	HOST ID

Fig. 9.12 Class C IP address format

28 bits	
1	1	1	0	MULTICAST GROUP ID

Fig. 9.13 Class D IP address format

9.6.1   Classless Inter-Domain Routing (CIDR) or Classless IPV4 Address

In a classless IP address, the Network Prefix is flexible represented by the following form,

$$X.Y.Y.Z / n$$

where n defines the number of bits for the Network Prefix. For example, the Class C IP address 192.10.20.1 uses 24 bits for the prefix and 8 bits for the host ID, which means that it can generate 256 IP addresses. This Class C IP address can be used as a CIDR address based on the number of Host IDs needed to determine prefix bits. If $n = 16$, then the IP address 192.10.10.1/16 has 16 bits for the network prefix and 16 bits for the host ID. Therefore, it can generate 2^{16} IP addresses where the starting address is 192.10.0.0/16 and the ending address is 192.10.255.255.

Assume a company needs 1000 IP addresses and only Class C addresses are available. This company is only able to use Class C addresses as classless addresses to assign the IPs to the interfaces.

CIDR Address Assignments

The IANA (Internet Assigned Numbers Authority) allocates a block of IP address to each ISP (Internet Service Provider). The ISP then distributes these IP addresses to their customers.

Assume a Network Information center allocates the block IP address 206.0.64.0/18 to Comcast. Comcast will have $2^{32-18} = 2^{14}$ IP addresses. If an organization requires 800 IP addresses, then Comcast will give the following addresses to the organization.

$2^{10} = 1024$ therefore, comcast allocates 10 bits of its host ID to this organization.

The organization address starts at 206.0.01000000.00000000 = 206.0.64.0/22 and ends at 206.0.01000011.11111111 = 206.0.67.

Subnetting and Classless Inter Domain Routing (CIDR)

In a classless IP address, the leading bits of the Host ID are used for Subnetting. The network prefix and new subnetting make the Extended network prefix.

Example 9.1 The classless IP address 169.194.168.82/27 is a host address.

(a) Find the number of addresses in the network.
(b) What are the starting and ending addresses?

    a. Since the Prefix is 27 bits and 32–27 = 5, there are 2^5 Host ID addresses.

       The starting address is:

          168.194.168.82 / 27 = 10100111.11000010.10100000.01010010

   where the least significant 5 bits are for the Host ID.

  b. The starting address is: 10100111.11000010.10100000.01**000000.**
       The last address is 10100111.11000010.10100000.01**011111.**

Example 9.2 An organization is granted IP addresses 15.24.74.0/24. The organization has three networks. Network A with 120 addresses, Network B with 60 addresses, and Network C with 32 addresses. Assign addresses to each network.

   The size of the Host ID is 32–24 = 8 bits. So, the total number of addresses available for the organization is $2^8 = 256$.

   Network A requires 120 addresses that use 7 bits for the host ID, where $2^7 = 128$ addresses.

   Network A's IP addresses start at 00001111.00011000.01001010.00000000/25.

   The ending address for Network A is 00001111.00011000.0100101 0.01111111/25.

   Network B requires 60 addresses, so it requires 6 bits, where $2^6 = 64$ addresses.

   The starting address of Network B is the next address after the ending address of Network A.

   The starting address of Network B is 00001111.00011000.0100101 0.10000000/26.

   The ending address of Network B is 00001111.00011000.01001010.10111111/26

.

   Network C requires 32 addresses, and it uses 5 bits for the Host ID. The starting address of Network C is the next address after the ending address of network B.

   The starting address of Network C is 00001111.00011000.0100101 0.11000000/27.

   The ending address of Network C is 00001111.00011000.01001010.11011111/27.

Loopback IP Address

The last address of each class is used as a loopback address for testing. The loop-back address is used on a computer to communicate with another process on the same computer.

The loopback addresses are:

Class A   127.0.0.1
Class B   191.255.0.0
Class C   223.255.255.0

Network Addresses

The host portion of network address is set to zero. For example, 129.49.0.0 is a network address, not a node address. No node is assigned to 0.0.

Broadcast Addresses

The host portion is set to all '1's in a broadcast. A packet with a broadcast address is sent to every node in the network. For example, address 129.49.255.255 is a **broadcast address**.

Private IP Addresses

The following IP addresses are reserved for 'private' networks

10.0.0.0   to   10.255.255.255
172.16.0.0   to   172.31.255.255
192.168.0.0   to   192.168.255.255
169.254.0.0   to   169.254.255.255.

9.7   Assigning IP Addresses to a Network

Most universities will have networks connected to the Internet, as shown in Fig. 9.14. The network administrator must contact InterNIC and obtain an IP address for the University. The InterNIC assigns a Class B address with a network ID of 129.47 (the first two bytes of the IP address). From these two bytes the network administrator must decide how many bits are needed for each subnetwork. This is determined by analyzing the growth of the network and the future needs of the university. In this example, we use one byte to represent our sub network address.

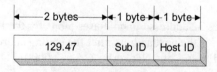

The subnetwork ID is 8 bits, implying that we can have 256 networks in the university with 255 nodes each. Each subnet ID is prefixed with the network ID. The following subnet IDs are assigned to each department as follows:

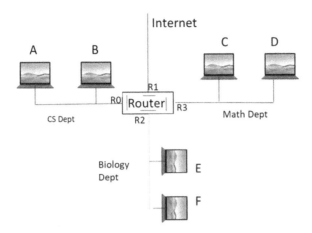

Fig. 9.14 A network with three LANs and one router

Subnet 1 (129.47.1.x) to the Computer Science Department.
Subnet 2 (129.47.2.x) to the Mathematics Department.
Subnet 3 (129.47.3.x) to the Biology Department.
Subnet 4 (129.47.4.x) to the link connected to the Internet.
Subnets 5 to 255 are unused and reserved for future use.

The network ID for the CS dept. is 129.47.1.0.
The network ID for the Math department is 129.47.2.0.
The network ID for the Bio department is 129.47.3.0.

In each subnet, there are 8 bits for host ID, meaning hosts can be assigned an address ending in the range 1–254.

IP addresses for the hosts on the CS department subnet are:

Node A's IP address: 129.47.1.1.
Node B's IP address: 129.47.1.2.

IP addresses for the hosts on the Math department subnet are:

Node C's IP address 129.47.2.1.
Node B's IP address 129.47.2.2.

IP addresses for the hosts on the Bio department subnet are:

Node A's IP address: 129.47.3.1.
Node B's IP address: 129.47.3.2.

In this example, the router has four ports, and each requires an IP address. Port R0, connected to CS Department, needs an IP on the same subnet, which would be 129.47.1.3. Similarly, port R1 is 129.47.4.1, as it is on the internet link subnet, and ports R2 and R3 are 129.47.3.3 and 129.47.2.3, respectively.

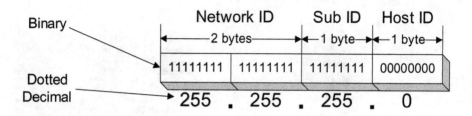

Fig. 9.15 Address mask

Address Mask

Address masks are used to define how many bits of an IP address are allocated for the host ID. It is used to separate the network address from a host ID. In the above example, the least significant byte is used for a host ID, as shown in Fig. 9.15.

9.8   TCP Connection and Disconnection

In a **TCP Connection**, the client first establishes a connection with the server before any information is transmitted. This connection process, commonly referred to as TCP three-way handshaking, is shown below in Fig. 9.16. The following steps describe the TCP connection set up.

1. The client sends a SYN packet to the server by setting the SYN flag to one and selecting a sequence number in TCP packet format (assume the sequence number is X).
2. The server responds to the Client by setting the SYN bit to 1, the ACK bit to 1, the acknowledge field to $X + 1$, and the sequence number to Y (sequence number of server). This packet is called a SYN-ACK packet.
3. The client then transmits the ACK packet with the ACK bit set to one, the Sequence number set to $X + 1$, and the ACK number (ACKnum) set to $Y + 1$.
4. Now that the connection has been established, the client sends 200 bytes of data to the server with sequence number $X + 1$, ACKnum $Y + 1$, and the ACK bit set to 1.
5. The server responds by sending 100 bytes of data back to the client, with the ACK bit set to 1 and the ACKnum set to $X + 201$. The ACKnum is the sum of the sequence number and the number of bytes received.
6. The client then responds with the ACK bit set to one, the sequence number $X + 201$, and the ACKNum $Y + 101$.

TCP Disconnection   When the source sends the last packet to the destination, the source sets FIN to 1 to inform the destination that this is the last packet. The destination acknowledges the last packet and sets the FIN to 1 to inform the source that the destination does not have any packet to send. The source sends a packet with RS set

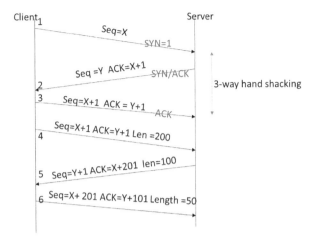

Fig. 9.16 TCP connection diagram

Fig. 9.17 Disconnecting a TCP connection between two hosts

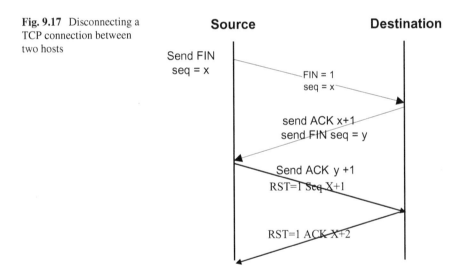

to 1 and the destination responds to the source with a packet whose RS bit is set to 1. Figure 9.17 shows the disconnection process.

9.9 ARP (Address Resolution Protocol)

The ARP command is used to display and modify entries in the Address Resolution Protocol (ARP) cache. The ARP cache contains the IP addresses and their resolved Ethernet or Token Ring physical addresses. Figure 9.18 shows a network with four computers, where IPi represents the IP address and MACi represents the MAC

Fig. 9.18 Network with
4 PCs

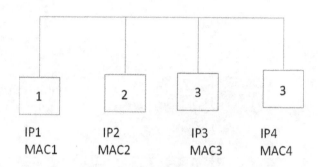

address of each PC. In order for PC1 to send a packet to PC4, PC1 requires MAC4. If the ARP cache of PC1 contains MAC4, then PC1 will be able to send a packet to PC4. If PC1 does not have MAC4 in its ARP cache, then PC1 will send an ARP command to request MAC4 of PC4. If PC1 wants to send a packet outside of the network, then it will send the packet to the default gateway.

ARP Commands

(a) Displays all ARP command options:

>   >arp

(b) Displays the ARP cache tables for all interfaces:

>   **>arp –a**

(c) Displays the ARP cache table for the interface that is assigned the IP address 10.0.0.99:

>   **>arp -a -N 10.0.0.99**

(d) Delete the IP address and corresponding MAC address from the ARP cache table:
>   >arp –d Internet address
>   a. Adding Internet address and MAC address to ARP cache table:

>   >arp –s Internet Address   MAC address
>   Intet_address: IP address represented by dotted decimal such as **10.12.10.23**
>   ether_addr: MAC address of the NIC represented in hexadecimal such as **cd-34-35-6f-ab-45**

>   b. Clear ARP cache table:

>   **>arp –d ***

ARP Packet Format

Figure 9.19 shows the ARP packet format. The following describes the function of each field in an ARP packet format:

0	15
Hardware Type 16 bits	
Protocol Type 16 bits	
HLEN Hardware address Length 8 bits	**PLEN IP address Length** 8 bits
Operation Code 16 bits ARP Request =1 ARP Response =2 RARP Request=3 RARP response =4	
Sender Hardware Address 48 bits	
Sender IP Address 32 bits	
Target Hardware Address 48 bits	
Target IP Address 32 bits	

Fig. 9.19 ARP packet format

Hardware Type

Hardware type identifies the type of hardware interface and following are some of the hardware types:

Type	Description
1	Ethernet
2	IEEE802.3
3	X.25
4	Token ring

Protocol type Protocol type identifies the type of protocol the sending device is using. For example: Protocol type 0800H is used for IP.

HELN Hardware Address Length (HELN) in bytes (means $6*8 = 48$ bits is the size of the hardware address).

PLEN Length of IP address.

Operation code Indicates whether the datagram is an ARP request or APR response.

$$1 = \text{ARP request}$$
$$2 = \text{ARP response}$$
$$3 = \text{RARP request}$$
$$4 = \text{RARP response.}$$

9.10   Demultiplexing Information

Figure 9.20 illustrates the general block diagram of Internet hardware and protocols. The packets (in the form of electrical signals) come to the physical layer of the Network Interface Card. The physical layer changes the signal to bits and passes it to the MAC sublayer. The MAC sublayer takes off its header (preamble, SFD, SA and DA) and passes it to the LLC (Logical Link Control) sublayer, which checks the type field. If the type field is 0800H the packet is an IPv4 datagram and is passed to IP.

For IPv6, the type field would be 86DDH (hexadecimal). IP looks at the 8-bit protocol field, removes its header, and passes the data to a protocol depending on the protocol number (TCP = 6, UDP = 17, ICMP = 1 and IGMP = 6). Assuming the data is passed to TCP, TCP will look at the port number and pass it to the application layer.

9.11   Internet Protocol Version 6 (IPv6)

Due to the growth of the Internet and the address limitations of IPv4, the Internet Engineering Task Force (IETF) approved IPv6 in 1995. The limitations that led to Internet Protocol Version 6 are summarized in the next paragraph.

The IPv4 address size is 32 bits and can connect up to 2^{32}, or 4 billion, users to the Internet. The IPv4 address field is divided into two parts: the network address (Network ID) and host address (Host ID). Once a network number is assigned to an organization, the organization might not use all host IDs in the host ID field, meaning that some IPv4 addresses may not be used. Also, the number of networks

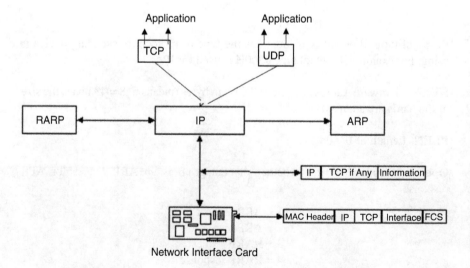

Fig. 9.20 Demultiplexing of information

connected to the exterior gateway increases rapidly which causes the routing table to become too large which ultimately increases the time it takes to search through the table.

The IPv6 protocol will reduce the size of the routing table in exterior gateways because IPv6 uses a hierarchical scheme to define an IP address. IPv6 has the following features:

- Expanded addressing
- Simplified header format
- Support extension
- Flow labeling
- Authentication and privacy

The IETF developed IPV6 and published the IPV6 document RFC 2460 in 1994. IPv6 has the following features:

1. Larger address space.

An IPV6 address is 128 bits which gives IPv6 an address range of 2^{128}, which is equivalent to
    340,282,366,920,938,463,463,374,607,431,770,000,000 addresses or
    340×10^{36}, as compared to the address space of IPV4, which results in only 2^{32}, or 4294967296, IPV4 addresses.

2. IPV6 contains less headers than IPV4.

Figure 9.21 compares the IPV4 packet format with IPV6.

IPV6 Header Explanation
Traffic class: Identifies different priorities.
Flow label: Used by the source to label the packets that require special handling by the routers such as real time service.

3. Hierarchical addressing (prefix).
    IPv6 uses 16 bits to represent the organization address.
4. Auto configuration.

    (a) **Stateless (RFC2462):** The host can automatically configure its IPv6 address based on the prefix advertised by router.
    (b) **Stateful (DHCPv6):** A DHCP server assigns an IPv6 address to the host.

5. Supports authentication and encryption.

    IPSec is mandatory for IPV6, but it is optional for IPV4.

6. Transition techniques to IPv4.

    (a) **Dual Stack**: A router that can accept and transmit IPv6 and IPV4 packets is called a Dual Stack router, as shown in Fig. 9.22.
    (b) **Tunneling**: Encapsulating IPv4 inside IPV6, as shown in Fig. 9.23.

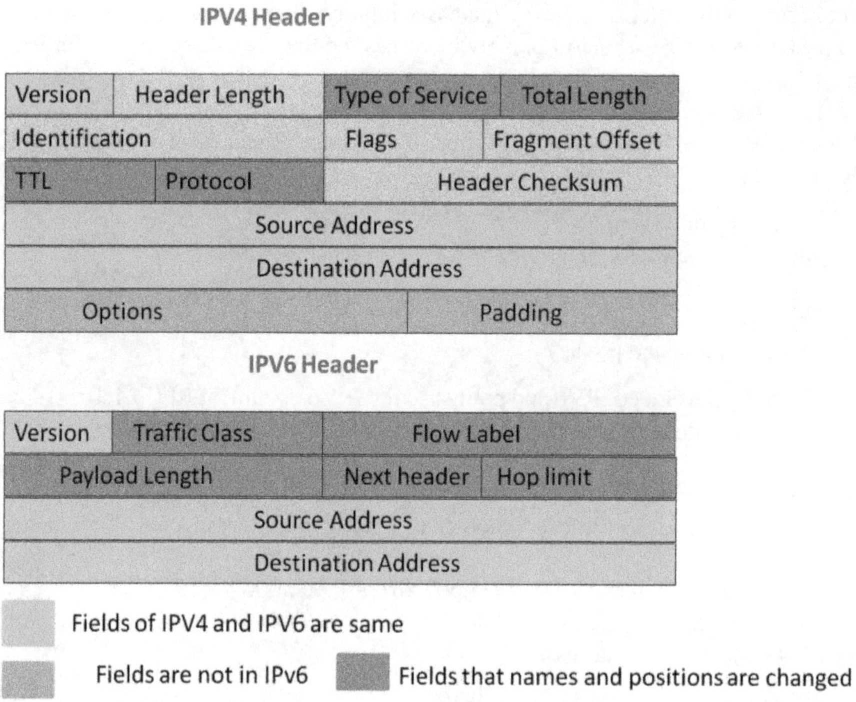

IPV4 Header
Version
Header Length
Type of Service
Total Length
Identification
Flags
Fragment Offset
TTL
Protocol
Header Checksum
Source Address
Destination Address
Options
Padding
IPV6 Header
Version
Traffic Class
Flow Label
Payload Length
Next header
Hop limit
Source Address
Destination Address
Fields of IPV4 and IPV6 are same
Fields are not in IPv6
Fields that names and positions are changed

Fig. 9.21 IPV6 and IPV4 Packet formats

Fig. 9.22 Dual Stack IPV6

Server
IPV6 & IPV4
Traffics
Dual stack
Router
IPV6
Internet
IPV4
Internet

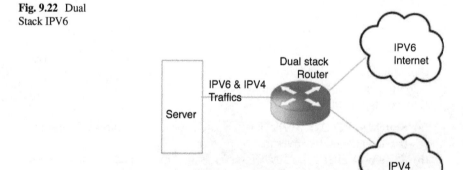

7. IPV6 does not require NAT (Network Address Translation).
8. Support for multihoming.

Multihoming allows an organization's network to be connected to two or more Internet Service Providers to offer high reliability.

Fig. 9.23 IPV4 encapsulation

9.12 IPV6 Address Architecture

The IPv6 address is 128 bits. It is divided into eight fields of 16 bits. Each field is represented in hexadecimal form:

$$Y:Y:Y:Y:Y:Y:Y:Y$$

Y is 4 digits in hexadecimal or 16 bits in binary.

Example 9.3
FE45:036A:00DF:4563:879E: 0008:0005:1232

The leading zero or zeroes of each field can be dropped; the above IPv6 address can be represented by

FE45:36A:DF:4563:8:5:1232

If an IPv6 address contains two or more contiguous fields of all zeros, these contiguous fields can be represented by a double colon; only one double colon in each address is allowed.

Example 9.4
FE45:436A:12DF:4563:879E: 8:0000:00000:1232 can be represented by
FE45:436A:12DF:4563:879E: 8**::**1232.

Example 9.5
The IPV6 address FE45:436A:0000:4563:879E:8:0000:00000:1232 can be represented by

FE45:436A:0:4563:879E: 8::1232.

IANA (Internet Assigned Number Authority) is responsible for global assignment of IP addresses. IPv6 addresses are generally assigned in a hierarchical manner.

IANA assigns a block of IPV6 addresses to the Regional Internet Register (RIR), then the RIR assigns a block of address to an ISP and the ISP assigns IPV6 addresses to organizations and users. The following are the list of Regional Internet Register (RIR).

- **AfriNIC**: African Network Information Centre for Africa
- **ARIN:** American Registry for Internet Numbers for the United States, Canada, Caribbean region, and Antarctica
- **APNIC**: Asia-Pacific Network Information Centre for Asia, Australia, and New Zealand.
- **LACNIC:** Latin America and Caribbean Network Information Center
- **RIPE NCC:** Réseaux IP Européens Network Coordination Centre for Europe, Russia, the Middle East, and Central Asia

9.13   IPV6 Address Format

The IPV6 address is 128 bits and Fig. 9.24 shows the IPV6 address format.

(RIR) Regional Internet Registry: 12 bits are assigned to a regional Internet Register.
(ISP) Internet Service Provider: The 20-bit ID of the ISP.
SITE: 16 bits used to identify an organization network.
Subnet: 16 bits used for organization network subnet.

The combination of RIR, ISP, and SITE is called the global prefix and it is 48 bits, as shown in Fig. 9.25.

9.14   IPV6 Address Types

IPV6 offers the following types of addresses.

1. Unicast address
2. Multicast address
3. Loopback address

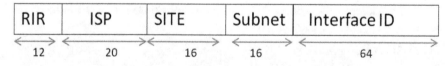

Fig. 9.24 Format of IPV6

Fig. 9.25 IPV6 general format

4. Anycast address
5. No broadcast address

Unicast: The unicast address defines addresses of a single interface that may have multiple IPv6 addresses. IPV6 unicast types are:

(a) Link-local address
(b) Site-local unicast address
(c) Global unicast address

Link-local address: A link local address is used by a node to communicate with other nodes that are on the same link. A router will not pass a link-local address to other links. Link-local addresses are identified by the prefix **FE80::/10**. A host will automatically configure its own local-link address. Figure 9.26 shows the link-local address format.

Interface ID: The 64-bit Interface ID can be configured by:

(a) EUI-64 (Extended Unique Identifier) bits using 48-bit MAC address
(b) Assign by DHCP
(c) Cryptography Generated Address

EUI-64: The 64-bit Extended Unique Identifier (EUI-64) was defined by the IEEE. It is generated by inserting the 16-bit value, 0xFFFE, into the middle of the unique 48-bit MAC address of the Network Interface Card (NIC) and inverting the MAC address's seventh bit from the left known as the universal/local bit. The resulting 64-bit Interface ID is shown in Fig. 9.27.

Example The MAC address of an NIC is **00 FE 01 12 AB CD**. By inserting FFFE into the middle of the MAC address and changing the U bit to one, the resulting 64-bit Interface ID is **02 FE 01 FF FE 12 AB CD**.

Site-Local Unicast Address
This type of address is used for local communication between nodes within a network and is not routable across the Internet. Figure 9.28 shows the format of a site local address with the prefix FEC0::/10.

Global Unicast Address
The global unicast address, as shown in Fig. 9.29, is equivalent to an IPV4 address and is used for global communication between nodes.

10 bits Prefix	54 bits	64 bits
1111 1110 10	0	Interface ID

FE80::/10

Fig. 9.26 IPV6 link-local format

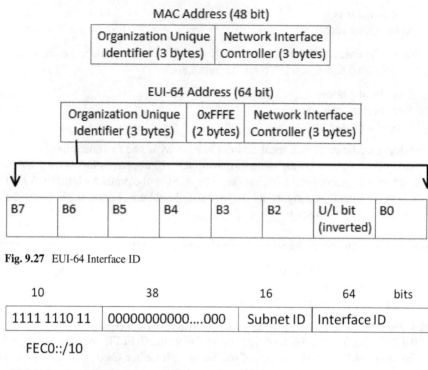

Fig. 9.27   EUI-64 Interface ID

10	38	16	64	bits
1111 1110 11	00000000000....000	Subnet ID	Interface ID	

FEC0::/10

Fig. 9.28   IPV6 site-local unicast format

Prefix (3 bits)	45 bits	16 bits	64 bits
001	Global Routing	Subnet ID	Interface ID

Fig. 9.29   Global unicast address format

Subnet ID: Defines the subnetwork ID within a site.

The prefix for a global unicast is 001. The IANA assigns a block of IPV6 unicast addresses to the following Regional Internet Registry (RIR).

- **AfriNIC**: African Network Information Centre for Africa.
- **ARIN**: American Registry for Internet Numbers for the United States, Canada,
- Caribbean region, and Antarctica.
- **APNIC**: Asia-Pacific Network Information Centre for Asia, Australia, and New Zealand.
- **LACNIC**: Latin America and Caribbean Network Information Center.
- **RIPE NCC**: Réseaux IP Européens Network Coordination Centre for Europe, Russia, and the Middle East, and Central Asia.

8bits	4 bits	4 bits	112 bits
1111 1111	Life time	Scope	Group ID

Fig. 9.30 IPV6 multicast format

IPv6 Multicast Address

IPV6 multicast is used for one-to-many delivery to all interfaces. Figure 9.30 shows the format of an IPv6 ulticast address. An IPV6 multicast address has the prefix FF00::/8.

Lifetime field: This field is represented by XXXT, where the XXX bits are unde-fined and T bit =1 means that the multicast address is temporary, and T = 0 means that the multicast address is permanent.

Scope field: Defines types of address:

Scope value	Types
0001	Interface local
0010	Link–local
0011	Subnet-local
0100	Admin local
0101	Site- local
1000	Organization local
1110	Global

Interface local: Used for transmission within a node.

Link-local multicast: Used for a local link and will not be forwarded to other links by devices on the network.

Site-local multicast: Used on single site.

Organization multicast address: This type of address is used within an organization.

Global multicast address: This type of address is used on the Internet and is simi-lar to a global unicast address. Some common uses of multicast addresses are:

FF02::1	All nodes in local-link
FF02::2	All routers on the local-link
FF05::2	All routers site-local
FF02::1:2	All DHCP agents link-local
FF05::1:3	All DHCP server site-local
FF05::FB	DNSv6

Solicited Multicast Address

Every node has a unicast address and special multicast address called the solicited mul-ticast address. The solicited multicast address of a node is represented by FF02:0:0.0.0:1:FFXX.XXX, where XX.XXXX is the 24-bit unicast address of the node.

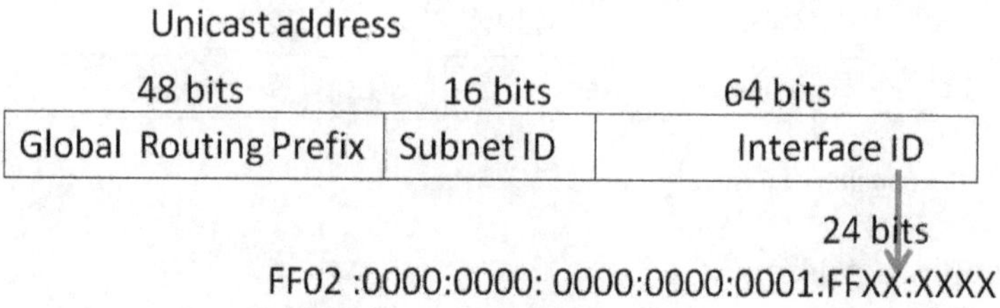

Unicast address

48 bits	16 bits	64 bits
Global Routing Prefix	Subnet ID	Interface ID

24 bits

FF02 :0000:0000: 0000:0000:0001:FFXX:XXXX

Fig. 9.31 Node-solicited multicast address generation

Example If the unicast address of a node is 2001.2400:0:0.2500:6A:2456:2CBA, then the solicited multicast address of the node will be FF02:0.0.0.1:FF56:2CBA.

Figure 9.31 shows node-solicited multicast address generation. The node-solicited address is used by network discovery protocol to request link layer address and Duplicate Address Detection (DAD).

Multicast Address for Routing Protocols

The following are some of the IPV6 multicast addresses for routing protocols:

FF02::5	OSPFv3 (OSPF IPv6) all routers
FF02::6	OSPFv3 designated routers
FF02::9	RIPng routers

Multicast Ethernet Address
An Ethernet address is 48 bits, and an Ethernet multicast address for IPV6 is represented by

33-33-XX-XX-XX-XX, where XX-XX-XX-XX are the least significant bytes of the IPV6 multicast address.

Example Convert the following IPV6 multicast address to a multicast Ethernet address.

Replacing xx-xx-xx-xx with FF-68-12-CB results in 33-33- *FF-68-12-CB*.

1. Loopback address

The address 0:0:0:0:0:0:1 is the IPV6 loopback address and is equivalent to the IPv4 127.0.0.1.

2. Anycast

An anycast packet is delivered to the nearest single interface, which is closest in terms of routing distance. The anycast address can be used by a client to request an IPV6 address from a server. Table 9.5 shows the prefix of an anycast address.

Table 9.5 IPV6 address prefix and types

Address type	Binary prefix		IPV6 notation
Unspecified	0000000……000	128 bits	::/128
Loopback	000000…….001	128 bits	::1/128
Link-local unicast	1111 1110 10		FF80::/10
Unique unicast local	1111 110		FC00::/7
Multicast	1111 1111		FF00::/8
Global unicast	001		2000::/3

9.15   IPv6 Addresses as URLs

IPv6 addresses can also be referenced in URLs (Uniform Resource Locator) by placing the IPV6 address in brackets.
http://[FEC0::CC1E:2412:1111:2222:3333]/index.htm

9.16   IPv6 Address Configuration

IPV6 addresses can be assigned to any host using Stateless or DHCPv6 Address Configuration.

Stateless Address Configuration
A node can automatically generate its IPv6 address without intervention or a DHCP Server by using one of the following methods to receive the 64-bit IPv6 prefix from a router.

A. Router advertisement: Router advertises its IPv6 on the link.
B. A node sends a Neighbor Discovery Packet (NDP) to all nodes on the link.
C. A node sends a solicitation request to the router, and the router responds with the request.

A node uses the IPv6 prefix to generate an IPV6 address through the following steps.

DHCPv6 (Stateful)
The Dynamic Host Configuration Protocol IPv6 (DHCPv6) server is used for stateful IPv6 configuration. DHCPv6 messages are:

DHCPv6 Solicit: This is an IP multicast message. The DHCPV6 of a client sends a DHCPv6 solicit message to FF02::1:2, which is the multicast address for all DHCPv6 (relays and servers). If received by a relay, the relay forwards the message to FF05::1:3, the multicast address of DHCPv6 servers.

DHCPv6 Advertise: This is a unicast message sent in response to a DHCPv6 Solicit. A DHCPv6 server will respond directly to the soliciting client if on the same link or through the relay agent if the DHCPv6 Solicit was forwarded by a relay.

DHCPv6 Request: After the client has located the DHCPv6 server, the DHCPv6 request (unicast message) is sent to request an IPV6 address from DHCPv6. The request must be forwarded by a relay if the server is not on the same link as the client.

DHCPv6 Reply: An IP unicast message sent in response to a DHCPv6 request which can be sent directly to the client or through a relay.

DHCPv6 Release: An IP unicast sent by the client to the server, informing the server of resources that are being released.

DHCPv6 Reconfigure: The DHCPv6 may send IP unicast or multicast messages to the client/clients for a new configuration and the client/clients must respond to the DHCPv6 request.

9.17   ICMPv6 and Neighbor Discovery Protocol (NDP)

ICMPv6 is a combination of ICMP and ARP with some additional functions. The functions of ICMPv6 are to generate error messages and information messages.

ICMPv6 Error Messages
(a) Packet is too large
(b) Path MTU discovery
(c) Destination unreachable
(d) Time exceeded

ICMPv6 Information Messages
(a) Ping command
(b) Echo request
(c) Echo response
(d) Neighbor Discovery Protocol
(e) Route Solicitation
(f) Route advertisement
(g) Neighbor solicitation
(h) Neighbor advertisement
(i) Multicast Listener Discovery
(j) Multicast listener query
(k) Multicast listener response
(l) Multicast listener done

Neighbor Discovery Protocol
Neighbor discovery protocol is used for:

1. Stateless Address configuration. The node solicits the router for the IPV6 prefix and the address of the default gateway.
2. Duplicate Address Detection (DAD).

3. Request the link layer address of neighbors (perform ARP function).
4. Determining if neighbors are reachable.

IPV6 Client Generation

On system startup, each node automatically generates a link-local address for its IPv6-enabled interface. The node uses NDP protocol to request an IPV6 prefix from the router (64 bits) and uses EUI-64 to generate an IPV6 address for its interface.

Summary

- The Internet is a collection of networks connected through gateways and routers.
- The Internet uses a set of protocols for communications called Transmission Control Protocol and Internet Protocol (TCP/IP).
- Some applications used by the Internet are E-mail, Telnet, File Transfer protocol (FTP), and Hyper Text Transfer protocol (HTTP).
- An e-mail address is made up of a user name and mail server, such as *Elahia1@ southernctedu*.
- Telnet is one of the most useful Internet applications. It enables one computer to remote login to another computer.
- File Transfer Protocol (FTP) is a protocol used for transferring a file from/to a remote computer without logging into the remote computer.
- Hyper Text Transfer Protocol (HTTP) is an advanced file retrieval protocol that can access distributed documents on the web.
- Transmission Control Protocol (TCP) provides reliable service between hosts.
- Internet Protocol (IP) provides packet delivery between the hosts.
- Internet Address Version 4 (IPv4) is 32 bits, where each byte is represented by a decimal number. They are divided into classes A, B, C, D and E.
- Class A IP address range is from 0.0.0.0 to 127.255.255. The Network IDs of 0 and 127 are reserved.
- The range of Class B IP address is from 128.0.0.0 to 191.255.255.255.
- The range of Class C IP address is from 192.0.0.0 to 223.255.255.
- The Class D IP addresses are reserved for multicasting and Class E IP addresses are reserved for future use.
- Domain Name System, or DNS, is a distributed database containing domain names and corresponding IP addresses. Top domain names are: edu, gov, com, mil, org, net, aero, coop, biz, name, and int.
- User Datagram Protocol (UDP) provides unreliable service between hosts. Applications of UDP are Trivial File Transfer Protocol (TFTP) and Remote Procedure Call (RCP).
- The TCP header is 20 bytes and the IP header is 20 bytes.
- IPv6 or IP next generation address size is 128 bits.

Key Terms

Address mask	Maximum Transfer Unit (MTU)
Address Resolution Protocol (ARP)	Multicast address
Advanced Research Project Agency (ARPA)	Network address
Anonymous FTP	NSFNET
ARPANET	Port number
Broadcast address	Remote login
Domain name	Telnet
Internet Protocol (IP)	Transmission Control Protocol (TCP)
Internet Protocol Version 6 (IPV6)	Unicast address
Loopback address	User Datagram Protocol (UDP)

Review Questions

Multiple Choice Questions

1. The Internet is a collection of _____.

   (a) servers
   (b) applications
   (c) networks
   (d) routers

2. The Internet uses the _____ protocol.

   (a) X.25
   (b) NWLink
   (c) TCP/IP
   (d) Window NT

3. Which one is not an application of the Internet?

   (a) E-mail
   (b) FTP
   (c) WWW
   (d) Antivirus software

4. _____ provides packet delivery between networks.

   (a) IP
   (b) TCP
   (c) X.25
   (d) ARP

5. The transport layer of TCP/IP consists of UDP and _____.

   (a) TCP
   (b) IP
   (c) a and b
   (d) ICMP

6. _____ performs reliable delivery of data.

   (a) IP
   (b) UDP
   (c) TCP
   (d) RARP

7. Which organization ratifies standards for the Internet?

   (a) IEEE
   (b) ITU
   (c) IETF
   (d) EIA

8. What type of switching is used in the Internet?

   (a) Virtual circuit
   (b) Packet switching
   (c) Circuit switching
   (d) Message switching

9. Which protocol is used for unreliable communication?

   (a) TCP
   (b) UDP
   (c) IP
   (d) ARP

10. Telnet uses which of the following protocols for remote login:

   (a) UDP
   (b) TCP
   (c) IP
   (d) FTP

11. Which protocol is used by a modem to connect a computer to the Internet?

   (a) TCP/IP
   (b) UDP
   (c) PPP
   (d) ARP

12. Telnet enables a user to_____.

 (a) transfer a file
 (b) send E-mail
 (c) remote login
 (d) transfer mail

13. How many bits are in an IPv4 address?

 (a) 24 bits
 (b) 32 bits
 (c) 48 bits
 (d) 128 bits

14. What is the application of a loopback address?

 (a) Reserved by Internet authority
 (b) Used for testing
 (c) Used for broadcast address
 (d) Used for unicast

15. Which of the following addresses is a Class C broadcast address?

 (a) 191.205.205.255
 (b) 191.205.205.205
 (c) 191.205.205.00
 (d) 192.205.255.2555

16. What protocol is used for the World Wide Web?

 (a) TCP/IP
 (b) HTPP
 (c) UDP
 (d) ARP

17. TCP is used for_____.

 (a) reliable communication
 (b) unreliable communication
 (c) connection oriented communication
 (d) none of the above

18. What is the function of the source and destination port in a TCP header?

 (a) It is used to identify the source and destination host on the network.
 (b) It is used to identify the application source protocol and application of des-
     tination protocol.
 (c) It is used to identify source protocol and destination protocol.
 (d) None of the above.

19. What is the function of Window size field in a TCP header?

    (a) The source reports its buffer size to the destination.
    (b) The source reports the number of packets received from the destination.
    (c) The source reports the size of its cache memory to the destination.
    (d) The source reports an error in the packet.

20. What is the function of Time-to-Live (TTL) in a TCP header?

    (a) It holds time of the day.
    (b) It defines the number of routers a datagram can pass.
    (c) It defines transmission time of a datagram between the source and destination.
    (d) It defines the number of words in a packet.

21. Which protocol is used to set up a connection between source and destination?

    (a) UDP
    (b) TCP
    (c) IP
    (d) ARP

22. How many bits is IPv6?

    (a) 32bits
    (b) 48 bits
    (c) 64 bits
    (d) 128 bits

23. The Internet routes a datagram from one gateway to another based on the datagram _____.

    (a) IP address
    (b) MAC address
    (c) Port address
    (d) None of the above

24. What is the function of an IP subnet mask?

    (a) IP subnet mask represents bits in the network portion of IP address.
    (b) IP subnet represents the bits in host portion of IP address.
    (c) a & b
    (d) None of the above

Short Answer Questions

1. What is the markup language used to create files for the WWW?
2. What is the protocol used by the WWW to transfer a file?
3. What is the protocol used to access the internet by a modem?
4. List the DNS indicators.
5. List the protocols in the transport level.

6. List the protocols in the Internet level.
7. What is the size of an IP_{v4} address?
8. What is the size of an IP header?
9. Explain the function of IP.
10. Explain the function of TCP.
11. What is the size of a TCP header?
12. What is the function of the TTL field in an IP header?
13. What is the size of a UDP header?
14. List the IP address classes.
15. If an organization uses the third byte of an IP address for a sub-net, how many sub-nets can be assigned to the Class B address?
16. What is the name of the organization applied to for IP addresses?
17. What is the function of ARP?
18. What is the newest version of IP?
19. What is the size of an IPv6 address?
20. Convert the following IP address from hexadecimal to dotted decimal representation, and find the class type of each IP address.

    (a) 46EF3A94
    (b) 23446FEC

21. List internet applications.
22. The following E-mail address is given; identify the username and mail server address:

    (a) Elahi@Xycorp.com

23. List two application protocols for UDP.
24. List three application protocols for TCP.
25. Show the TCP/IP Reference Model.
26. What is the function of ICMP?
27. What does DNS stand for? Describe its function.
28. Explain the function of telnet.
29. What is a port number?
30. What is an MTU?
31. Identify the class of the following IP addresses:

    (a) 129.234.12.08
    (b) 117.243.56.89
    (c) 92.92.92.9

32. How many bits is IPv6?
33. Assume the Internet Network Information Center assigns you a Class B IP address 172.200.0.0.

    (a) How many bits do you use from the host ID for a 128 subnet ID?
    (b) How many host IDs can be generated for each subnet ID?
    (c) What is the subnet Mask ID?

34. There are two computers, A and B, with IP addresses of 174.20.45.37 and 174. 20.67.45.

If these two computers have a subnet mask ID of 255.255.0.0, can you determine if these two computers are in the same network?

35. The following figure shows the network of an organization. As a network administrator:

(a) Assign an IP address to each network and interface using a Class B address.
(b) Show the routing table for each router.

Chapter 10
Internet Protocols Part II and MPLS

Objectives

After completing this chapter, you should be able to:

- Explain the application of DNS.
- List DNS operations.
- List top level DNS domains.
- Define the components of DHCP.
- Explain the operation of DHCP.
- Explain the operation of HTTP.
- Define the types of HTTP packets.
- Explain Internet Control Message Protocol (ICMP).
- List several routing protocols.
- Explain link characteristics.
- List network diagnostic commands.
- Explain the operation of MPLS.
- List the applications of IP multicasting.
- Able to write simple socket program.

10.1   Domain Name System (DNS)

Introduction

Memorizing a word is much easier than memorizing a long number, such as seen when trying to remember the word "Yahoo" vs the number "67.195.160.76". Therefore, computer hostnames are frequently represented by a word rather than an IP address. In order for a client computer to access a website, send an email, or connect to another computer through the Internet, the client computer must have the IP address of the destination, but this may not be known to the user. To accomplish this,

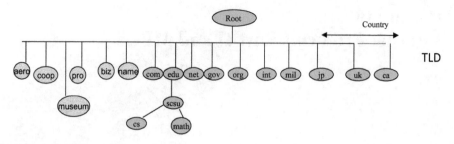

Fig. 10.1 DNS top level domain names

the client computer uses domain name system (DNS) to obtain resolve a known hostname to the IP address of the host. The RFC 881, 882, and 883, which have been published by IETF, explain DNS operation.

DNS is a distributed database containing information about domain names and their corresponding IP addresses. DNS uses a tree hierarchy that consists of a root and top level domains (TLDs). The TLD defines the type of organization, such as "edu," which represents an educational organization. Top level domain names can be categorized as shown in Fig. 10.1.

aero	Air transport industry	museum	for museums
edu	Educational sites	**mil**	Military
gov	Government sites	**org**	Organization (nonprofit organization)
com	Commercial sites	**int**	International Organization
net	Major Network Providers	**biz**	for businesses
coop	cooperatives	**name**	use for individual

A two-letter abbreviation is used for a particular country, such as "uk" for the United Kingdom and "fr" for France.

An organization must register its sub-domain name with the Internet Assigned Number Authority (IANA). For instance, southernct.edu is registered by IANA and must be a unique name in its domain. Southern must also specify two DNS servers: primary and secondary, which hold the hosts' names and corresponding IP addresses.

DNS Root Servers
There are 13 DNS root servers, and they support both IPv4 and IPv6; the website https://www.iana.org/domains/root/servers shows information about DNS root servers.

DNS Names
DNS names use a fully qualified domain name (FQDN). A FQDN is a DNS domain name made of its location on the TLD separated by ".", such as cs.southernct.edu.

DNS Operation
The function of DNS is to convert the host name to the IP address much like a phone directory converts a name to a phone number. In order for a client computer to

access the yahoo.com web site, the client must have the IP address of the Yahoo server. The client must send a query to the DNS server and request the IP address of Yahoo. There are two methods that can be used by a client to request an IP address. These methods are DNS recursive queries and DNS iterative queries. By default, DNS uses iterative queries.

A. DNS Iterative Query

The client computer at southern requests connection to yahoo.com. The client computer requires the IP address of yahoo.com's server. The following steps describe the iterative query that allows the client to obtain the IP address of yahoo as shown in Fig. 10.2.

1. The client sends a query to the southernct.edu DNS server (ns1.southernct.edu) to resolve the name to an IP address. If Southern's DNS server resolves the client request, then it will send the IP address of yahoo's server to the client. If Southern's sever does not have IP address for yahoo, the process goes to step 2.
2. Southern's DNS server sends a request to the Root DNS server, requesting the IP address of yahoo.
3. The Root server sees that the requested information (FDQN) name has a TLD with **.com**. It sends the IP address of one of the .com servers that has the IP address of Yahoo's DNS server to Southern's DNS.
4. Southern's server sends the request to the .com TLD server to resolve the name to an IP address.

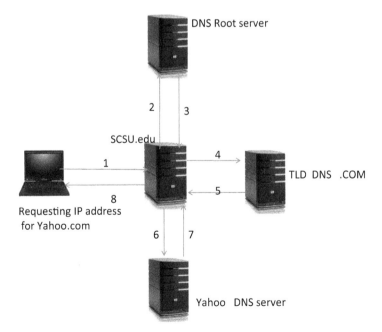

Fig. 10.2 Iterative DNS query

5. The .com TLD server responds with the IP address of yahoo's DNS server.
6. Southern's DNS server sends a request to the DNS server of yahoo (authoritative) for the IP address.
7. The Yahoo DNS server resolves the request and transmits the IP address to Southern's Server.
8. Southern's server transmits the IP address to the client.

B. Recursive Query Operation

In a recursive query, the root DNS obtains the IP address requested by the client and sends it to the client, as shown in Fig. 10.3.

Primary and Secondary Domain Name Servers
The function of secondary DNS server is to provide back up for primary DNS server. If the primary server goes down, then clients of an organization cannot resolve the host name to the IP address without having a backup server. A primary domain name server holds the zone files (resource records) used for resolving host names to IP addresses.

Zones and Zone Files
Zones

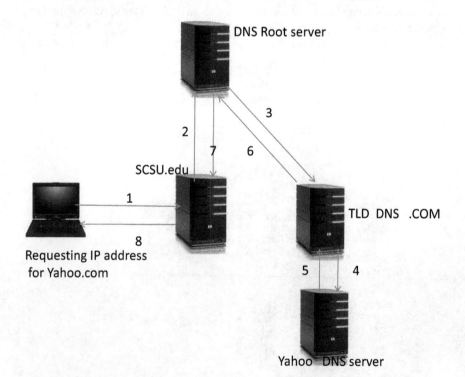

Fig. 10.3 Recursive query operation

Fig. 10.4 SouthernCT
DNS zone

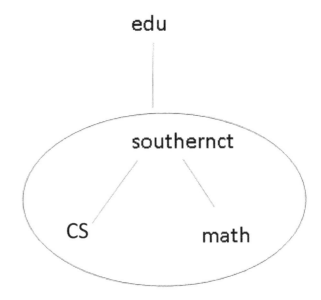

A zone is a part of domain name system that contains resource records and is managed by an administrator. Figure 10.4 shows one of Southern's DNS zones. The southernct.edu can have one zone which include all subdomains, or split each subdomain into separate zones or across DNS servers.

Zone file

The zone file holds resource records (information about domain names and their corresponding IP addresses). This file is transferred to the secondary server using TCP connection.

Resolver
A resolver is a program that runs on a client for generating query messages from the server, and the same program runs on a DNS server for responding to DNS queries.

C. DNS Packet Format

DNS uses the UDP protocol for transmitting query packets and receiving query responses. Figure 10.5 shows the DNS packet format.

Identification (16 bits): The identification field is used to identify the packet, and it is generated by the device transmitting the query packet. The response packet uses the same identification number as the query packet.
Flags: The flag field is 16 bits, and contains the following flags:

1	4	1	1	1	1	3	4
QR	Operation Code	AA	TC	RD	RA	Reserved	Response code

0	31

Identification	Flags
Number of Questions	Number of answer RRs
Number of Authority	Number of additional RRs
Questions (Name, Type)	
Answers (Resource Record)	
Authority	
Additional Information	

Fig. 10.5 DNS packet format

QR (Query/Response): QR =0 means that a packet is query packet, and QR = 1 indicates the packet is a response to a query.

Operation Code: The operation code defines the type of query message, and can be the following:

Operation code	Description
0	Standard query
1	Reverse query
2	Request server status
3	Reserved
4	Primary server informing the secondary server that data for a zone has changed

AA (Authoritative Answer Flag): This bit is set to 1 in a response to indicate that the server that created the response is authoritative for the zone. If it is 0, that means that the response is from a nonauthoritative source.

TC (truncated): This flag set to 1 means that the message was truncated, while 0 indicates otherwise.

RD: RD = 1 means that the client has requested a recursion method.

RA: This flag is used as response to RD, and when set to 1 means that the response uses recursion.

Response Code: Some of the response codes are:

Code value	Description
0	No error
1	Format error
2	Server was unable to respond
3	Name in the query does not exist in the RRs

Questions: Defines the number of questions in the query message.

Answers: Defines the number of answers in the query response.

Authority: Specifies the number of resource records in the *Authority* section of the message.

Number of additional RRs: Defines the number of RRs in the additional section.

D. DNS Resource Record

The resource records contains the following information:

Name	Class	Type	Value	TTL
yahoo.com	IN	A	209.191.122.70	3600 s (1 h)

1. Name: The domain name that the resource record belongs to (the domain name may have more than one resource record).
2. Time-to-Live: This is a 32-bit integer. The TTL is measured in seconds. The value zero indicates the data should not be cached.
3. Class: This field usually contains the value 'IN', and it represents if this record is to be used by the internet.
4. Type: The type field defines the type of resource record, which can be:

Type	Description
A	The address in the record is IPV4.
AAAA	The address in the record is IPv6.
MX	The record is a mail exchange.
NS	The record is a name server.
PTR	This is a pointer that points to another file to resolve name server and IP.
CNAME	This is canonical name type of record.
SOA	This is the start of authority type of record.

5. Value: This field can be a number, ASCII strings, or any domain. The semantics of name and value depends on the type field.

DNS Root Servers

DNS root servers are a network of 100 servers that are located in different countries and managed by 12 organizations. For locations of the root servers, visit https://root-servers.org/.

10.2   Dynamic Host Configuration Protocol (DHCP)

Introduction

Manually assigning IP addresses to host computers in a large network is very time consuming. To overcome this problem, TCP/IP offers Dynamic Host Configuration

Protocol (DHCP), which is an extension of the boot protocol. DHCP uses the UDP protocol to communicate with a DHCP server. The client broadcasts a packet with the IP address 255.255.255.255 (broadcast address). The broadcasted packet contains the hardware address of the client. The DHCP sever responds to this request.

DHCP Components

To have a DHCP service on a network, it requires three types of software: a DHCP client, DHCP server, and DHCP relay agent.

DHCP Client Most Network Operating Systems (NOS) offer DHCP software for clients such as Windows 10. The DHCP client software enables the client workstation to obtain its IP address from the DHCP server automatically. Clients broadcast a packet to the network that has the DHCP server in its broadcast domain (i.e., the same segment).

DHCP Server The DHCP server holds a range of IP addresses and responds to any request made by a DHCP client. Note that the DHCP server and the client host must be in the same broadcast domain; otherwise, the DHCP relay agent software is required.

DHCP Relay Agent When the DHCP server is not located in a broadcast domain of the client station, the router that the client station is connected to requires the DHCP relay agent software, as shown in Fig. 10.6.

The function of the DHCP relay agent is to accept the broadcast packet from the DHCP client and send the packet to the DHCP server. The relay agent changes the broadcast packet to a unicast address by using its own IP address, then sends it to the DHCP server. The DHCP server responds to the DHCP relay agent. The DHCP relay agent then forwards this response to the DHCP client.

The DHCP server supports different methods to allocate IP addresses to a client host:

1. **Automatic allocation:** DHCP assigns a permanent IP address to the host.
2. **Dynamic allocation:** DHCP assigns an IP address to a host for a limited period of time (this time is called lease time). If the client does not need the IP address, the DHCP server can reuse this IP address and assign it to another host. This method is used by internet service providers to assign IP addresses to their clients in order to be connected to the Internet temporarily.
3. **Manual allocation:** The network administrator assigns an IP address to the host and the DHCP server transfers that IP address to the client.

Fig. 10.6 A router with a relay agent in a client host/DHCP server connection

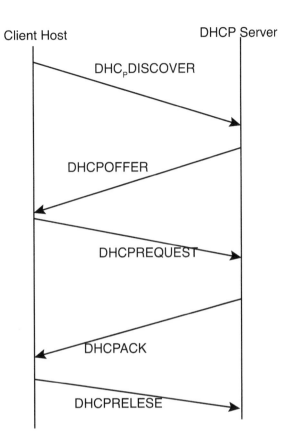

Fig. 10.7 DHCP operation

DHCP Operation
The following steps describe the operation of DHCP, as shown in Fig. 10.7:

1. The DHCP client does not have an IP address, and so it broadcasts a packet on the network requesting an IP address from the DHCP server. This packet is called a **DHCP Discovery** packet.
2. The DHCP server responds to the DHCP Discovery packet by sending a **DHCP Offer** packet to the client. The DHCP Offer packet includes the client IP address, the address mask and the lease time (the amount of the time that the client can hold this address). In the middle of the lease time the client host sends a renewal packet to the DHCP server to find out if it can keep the same IP address for the next lease time.
3. When the client host has received a DHCP Offer packet from the DHCP server, it can accept the address offered or reject it. Therefore, the client sends a packet to the DHCP server to inform either the acceptance or rejection of the IP address. This packet is called the DHCP Request packet.
4. The DHCP server sends a DHCP ACK to the client in response to the client's DHCP Request and informs the client of the completion of the DHCP process.

5. The client receives the DHCP ACK packet with configuration from the DHCP server. Now the client is configured. If a client receives a DHCP NACK (negative acknowledgment from the DHCP server), then the client host cannot use this IP address.
6. If the client host does not need an IP address, it will send a DHCP Release packet to the DHCP server to release the IP address.

DHCP Packet Format

DHCP is an application protocol for UDP. Figure 10.8 shows the payload of an IP packet with DHCP. Figure 10.9 shows the DHCP packet format.

IP header	UDP header	DHCP packet

The following describes the function of each field in the DHCP packet format:

OP code: The Op code indicates a request from a client or a reply to a request (1 for request, 2 for reply).

Hardware type: The hardware type indicates the type of network card being used (such as IEEE 802.3 or Token Ring).

Hardware length: The hardware length indicates the size of the hardware address or MAC address (6 bytes).

Hops: The hops field indicates the number of hops a packet can make on route to the destination. The maximum number of hops is 3.

Transaction ID: A random number set by the client. It is used by the client and the server to coordinate messages and responses.

IP Header	UDP header	DHCP Packet

Fig. 10.8 IP packet with DHCP

1   32

Op Code 8 bits	Hardware Type 8 bits	Hardware Length 8 bits	Hop Count 8 bits
Transaction ID 32 bits			
Number of Seconds 16 bits		Unused 16 bits	
Client IP Address 4 bytes			
Machine IP Address 4 bytes			
Server IP Address 4 bytes			
Gateway IP Address 4 bytes			
Client MAC Address 6 bytes			
Server Host Name up to 64 bytes			
Boot file name up to 1284 bytes			
Vendor-Specification Information up to 64 bytes			

Fig. 10.9 DHCP packet format

Number of seconds: The number of seconds is set by the client. The secondary
   server does not respond until this time has expired.
Client IP address: If the client does not have an IP address, this field will be set to
   0.0.0.0.
Server IP address: The server IP address is set by the server.
Router IP: The router IP is set by the forwarding router.
Client hardware: The client address is set by the client and is used by the server to
   identify which client the request came from.
Server host name: optional.
Boot file name: The client can leave this field null or indicate the type of the
   boot file.
Vendor specification: This field is used for various extensions of the bootstrap.
UDP header: The UDP header contains source and destination port numbers. The
   BOOTP uses two reserved port numbers. Port number 68 is used for the client
   and port number 67 is used for the server.

10.3   HTTP (Hypertext Transfer Protocol)

The IETF published HTTP (Hypertext Transfer Protocol0) documents RFC 1945
and RFC 2616. HTTP is an application protocol for TCP that uses port 80 for communications between clients and servers. Hypertext Transfer Protocol is used for the
transferring of Hypertext Markup Language (HTML) documents. HTML is a tag,
and hypertext documents have links to other documents such as images and videos.

   The address of a web page is defined by a URL (*Uniform Resource Locator*),
such as http://www.southernct.edu/csdept/picture.gif

   Where http is the protocol, southern is the host name, ".edu" is the TLD, and
csdept/picture.gif is the path name.

   In HTTP, the client requests a web page and the server responds by sending the
web page to the client, as shown in Fig. 10.10.

   HTTP is implemented by both client and server software. It performs the following tasks for transferring a web page from a server to a requesting client.

1. The browser makes a connection to the server over TCP port 80.
2. The browser sends a request to the server.
3. The server sends a response to the browser.

Fig. 10.10 HTTP
operation

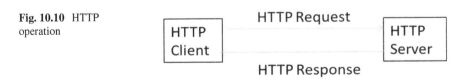

HTTP Characteristics

1. HTTP is a stateless protocol, meaning that an http server does not keep any information about the clients that requested web page.
2. HTTP connections: HTTP offers two type of connections called nonpersistent and persistent connections.

   A. **Nonpersistent connection:** Each request and response are sent over a separate connection and the connection is closed after a single request/response pair. HTTP/0.9 and HTTP 1.0 use non-persistent connection.
   B. Persistent connection: All requests and responses are sent over the same connection. HTTP version 1.1 uses persistent connection.

HTTP Packet Format

HTTP defines two types of packets: client packets and server packets.

A. **HTTP Client Packet Format**: Figure 10.11 shows an HTTP packet format that consists of a Request line, Header lines, a Blank line, and a Message Body.

SP: Space
CR: Carriage Return
LF: Line Feed
CR = %0d = \r
LF = %0a = \n
% means hex

Method: Defines the type of request sent by the client to the server, such as:

GET: Used by the client to request a document from the server and is identified by the URL.
HEAD: The HEAD method is similar to GET except that the server does not return a message-body in the response.
POST: Used by a client to submit information to the server.
PUT: Used for updating information on the server.

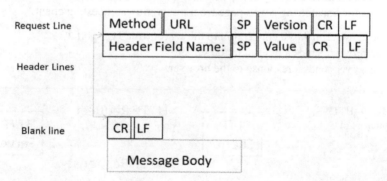

Fig. 10.11 HTTP request packet format

Example of a Client Request
Request line   **GET   /csdept/picture.gif   HTTP/1.1**
Header line   **HOST: www.southernct.edu**
Header line   **Connection: close**   ; uses non-persistent connection
Header Line   **User-Agent: Internet Explorer**
Header Line   **Accepted-Language: English**

HTTP Response Packet Format
An HTTP server responds to a client request with an HTTP response packet.
Figure 10.12 shows HTTP packet format.
     The following describes the function of each filed in Fig. 10.12.

Status line   **HTTP 1.1   200   ok**
Header Line   **Connection:**   close
Header Line   **Server**: csdept
Header Line   **Last-Modified:** Mon, 6, 2010 09:21GMT
Header Line   **Content-Length:** 500
Header Line   **Content-Type:** Text/HTML
Blank line
Message Body

Status Code Definitions
Success (2XX): This class of codes indicates that request was successful, which
includes:

200 OK
201 Created
202 Accepted
203 Non-Authoritative Information
204 No Content −205 Reset Content
206 Partial Content

     Redirection (3xx): This class of codes indicates that an action is required to be
taken by the client in order to complete the request.

301 Moved Permanently
302 Moved Temporarily

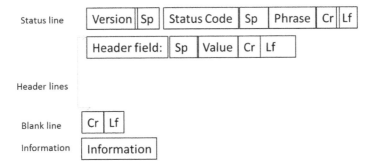

Fig. 10.12 HTTP Response packet format

Client Error (4xx): This class of codes indicates client errors such as:

400 Bad Request
401 Unauthorized
402 Payment Required
403 Forbidden
404 Not Found.

10.4   Internet Control Message Protocol

Internet Control Message Protocol (ICMP) is used to report error messages during packet delivery by an Internet Protocol such as protocol unreachable, network unreachable, network congestion, packet is too large, and announcing timeouts (TTL field drops to zero). Also, ICMP is used by the network administration for diagnostics purposes, such as the *Ping* and *Tracert* commands. ICMP is an Internet Layer Protocol and ICMP packets are transported by IP to the destination. ICMP messages can be divided into two groups: error messages and information messages. Figure 10.10 shows the ICMP packet format (Fig. 10.13).

Type (8 bits): Indicates the type of message. Numbers 0 through 127 indicate error messages, and numbers larger than 127 represent information messages. Table 10.1 shows type values and their meanings.

Code (8 bits): Provides additional information about the message type. For example, for a type value of 1 (destination unreachable), the code number specifies more detail information as shown in Table 10.2.

Checksum: This field is used for error detection of ICMP headers and messages.

Packet too large: Consider Fig. 10.14, in which two networks are connected by a router. Network A is a token ring with MTU of 18,000 bytes and Network B is an Ethernet with MTU of 1500 bytes. If station A sends an IP packet to station B, station A sends the packet to the router and the router examines the DF bit (do not fragment) in the IP header. If this bit is set, indicating that the packet is not to be fragmented, then the router is not able to send the packet to station B and the router discards the packet. The router then sends an error message to station A indicating that the packet is too large and the MTU for the packet should be 1500

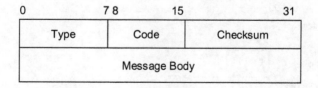

Fig. 10.13 ICMP packet format

Table 10.1 Type values and meaning

Type value	Meaning
1	Destination unreachable
2	Packet is too large
3	Time exceeded
4	Parameter problem
128	Echo request
129	Echo response

Table 10.2 Error messages and associated code values

Code	Definition
0	No route to destination
1	Communication with destination prohibited
2	No route to destination
3	Not assigned
4	Address unreachable
129	Port unreachable

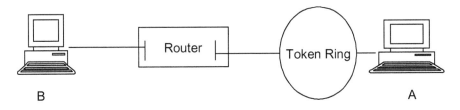

Fig. 10.14 Two networks with different MTUs are connected via a router

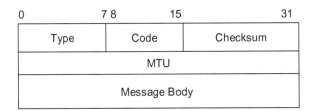

Fig. 10.15 ICMP packet format for MTU discovery

bytes. Station A sends a new packet with the MTU of 1500 bytes to the router. This is called *Path MTU discovery*. Figure 10.15 shows the ICMP packet format for Path MTU discovery.

In Fig. 10.15, the type field is equal to 2, code field is zero, and MTU field is maximum transmission unit of the next-hop link.

Time exceeded: When an IP packet travels through the Internet, the TTL field dec-
rements by one each time the packet passes through a gateway. When the TTL
value becomes zero, the gateway sends the ICMP message of Time Exceeded.

Parameter problem: When a router discovers an error in an IP packet, the router
discards the packet and sends a Parameter Problem to the source.

Echo request and response: An ICMP echo-request message, which is generated
by the *ping* command, is sent by any host to test node reachability across an
Internetwork. The ICMP echo-reply message indicates that the node can be successfully reached.

10.5   Routing

The function of a router is to determine the path for transporting information (packets) through the Internet. Furthermore, routing is the method of moving packets
from one network to another network. Figure 10.16 shows four networks connected
through the routers R1, R2, R3, and R4. Consider the following questions when host
A wants to send a packet to host B:

1. How does host A know that host B is connected to router R3?
2. What is the best route for sending information from host A to host B? Host A can
    send packets to host B by using three different paths. Host A can send packets to
    host B using the path R1-R2-R3 or R1-R4-R3 or R1-R2- R4-R3. The router
    determines the best route for sending information by using a routing algorithm
    to build a routing table.

Link Characteristics
Routers use different metrics (link characteristics) to find the best route for sending
a packet to the destination. The metrics are:

Hop count: The hop count measures the number of hops that a packet must go
through in order to get to the destination.

Throughput: The data rate of the link.

Communication cost: The cost of transmitting information from source to
destination.

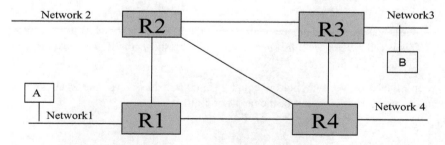

Fig. 10.16 Networks connected by four routers

Delay: Measures the amount of time it takes a packet to travel from source to destination.

Configuring the Routing Table
There are two ways to configure the routing table: **dynamic routing** and **static routing**.

Dynamic routing: In dynamic routing, the routers use a routing protocol to build their routing tables. If there is a change in the network configuration, a dynamic routing protocol broadcasts the changes to all the routers in the network in order to update all routing tables. Some of the most popular dynamic routing protocols are:

(a) Routing Information Protocol (RIP)
(b) Open Short Path First Protocol (OSPF)
(c) Interior Gateway Routing Protocol (IGRP)

Static routing: Static routing tables are configured manually by the network administrator. Static routing is often used for small networks. The problem with this type of routing is that if there is a change in network topology or a network link failure, all the routing tables need to be manually updated.

Figure 10.17 shows a network with three routers: A, B, and C. Router A has an Ethernet link E0 and one serial link S0. Router B has two serial links S0 and S1 and one Ethernet link E0. Router C has one serial link S0 and one Ethernet link E0. Class B IP addresses (180.160.0.0) are used to assign an IP address to each host.

As a network administrator, one would need to assign an IP address to each network and to each host.

The following IP addresses are assigned to the following networks:

180.160.10.   the network connected to the Ethernet port of router A
180.160.20.   the serial link between router A and router B
180.160.30.   the Ethernet network connected to router B
180.160.40.   the serial link between routers B and C
180.160.50.   the Ethernet network connected to router C

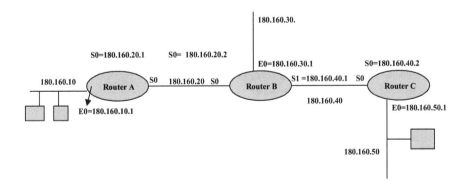

Fig. 10.17 Networks connected by three routers

Table 10.3 Routing table for
router A

Remote network	Subnet mask	Default gateway
180.160.50.	255.255.255	180.160.20.2
180.160.40.	255.255.255	180.160.20.2
180.160.30.	255.255.255	180.160.20.2

Table 10.4 Routing table for router B

Remote network	Subnet mask	Default gateway
180.160.50.	255.255.255.	180.160.40.2 (S0 of router C)
180.160.10	255.255.255.	180.160.20.1 (S0 of router A)

Table 10.5 Routing table for router C

Remote network	Subnet mask	Default gateway
180.160.10	255.255.255.	180.160.40.1 (S1 of router B)
180.160.120	255.255.255.	180.160.40.1 (S1 of router B)
180.160.30.	255.255.255.	180.160.40.1 (S1 of router B)

In static routing, the routing table is built and updated manually. The IP routing table has the following fields:

A. IP address of remote network
B. Subnet mask
C. IP address of default gateway

Tables 10.3, 10.4, and 10.5 show the routing table for routers A, B, and C.
Router A can access the other networks through the S0 link of router B. Therefore, S0 is the default gateway for router A.

10.6   Networking Diagnostic Commands

TCP/IP developed networking commands for network diagnostics, such as **ping**, **arp, ipconfig, tracert, netstat**, and **nslookup**. The following commands are based on the Microsoft Windows OS:

A. Ping command

The **Ping (Packet Internet Groper) command** is one of the most useful tools for debugging networks. It is an ICMP echo process used for checking whether a destination host is reachable. Ping transmits a 32-byte ICMP echo request to the destination and expects to receive a 32-byte ICMP echo packet from the destination. It also displays the time it took to transmit 32 bytes and receive 32 bytes of data.

B. Address Resolution Protocol

If a router receives a packet for which it does not have the hardware address of the packet in its Address Resolution Protocol (arp) cache table, the router sends a broadcast packet to the hosts on the network and requests the hardware address of the host. All the hosts in the network accept the packet and compare the IP address of the packet with its own IP address. If both IP addresses are the same, the host will respond to the router with a RARP, which contains the hardware address of the host.

arp tables contain both the IP address and the MAC address of a computer. The following is a list of arp commands:

arp -a   Display the contents of the ARP table
arp -d   Delete an entry with an IP address
arp -s   Add an entry with a MAC address

C. ipconfig /all

ipconfig displays the network settings, physical address, IP address, and subnet mask of the host, as well as the IP address of the default gateway.

D. Tracert command

Tracert shows the path of a packet from source to destination and the number of gateways the packet travels through.

E. Netstat command

The netstat command displays information about your network configuration. It comes with following options:

netstat –n   displays information on the NIC of your computer
netstat –r   displays the IP routing table
netstat –a   displays information on TCP and UDP ports
netstat –s   displays operational statistics of network protocols

10.7   Multi-Protocol Label Switching (MPLS)

MPLS was developed to overcome the deficiency of IP packet delivery over private and public networks. An IP network is connectionless and when an IP packet is received in a router, the router uses its routing table and the IP address of the packet to find the next hop. The IP address does not use a fixed size, such as class A, B, C, or D; therefore, it takes time to find the next hop. MPLS defines a method for fast packet forwarding over an IP network. It is an independent protocol, and it works with multiple protocols, such as IP, ATM, and Frame Relay. MPLS adds a label or tag to an IP packet, and it uses this label to find the path or next hop. Because the length of the label is fixed, it becomes faster to look up in a table than IP routing. Routers on an IP network use this label to find the next hop. The operation of MPLS is similar to cell switching in ATM networks and Frame Relay. MPLS performs the same function as a router but with higher performance. MPLS can be used for VPN, transport layer (layer 2), and connection-oriented service.

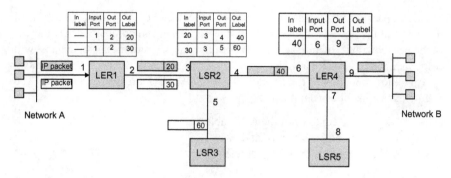

Fig. 10.18 MPLS network

MPLS Components and Operation: Figure 10.18 shows an MPLS network that consist of the following components:

A. *Label Edge Router (LER)*: An LER is located between the MPLS network and the user networks, as shown in Fig. 10.18. An LER performs the following functions:

- When an LER receives an IP packet from a user network, it adds its label to the IP packet and uses its routing table to forward the IP packet based on the label value to the next label switch router.
- When an LER wants to transmit a packet to the user network, it removes the label field from the IP address and transmits the IP packet to the user network.

B. *Label Switch Router (LSR)*: The function of an LSR is to examine the IP packet based on its label and replace the label with a new label. The new label identifies the next hop to which the packet is forwarded.

In Fig. 10.18, network A transmits two IP packets to LER1. LER1 uses its routing table to add labels to each packet and forward them to LSR2. LSR2 examines each packet's label and strips the packet of its label, then inserts a new label, and forwards the packet to the appropriate port of the next hop. IP packet with label 40 is transmitted to LER 4 and IP packet with label 60 is transmitted to LSR3. LER4 examines the label of the IP packet and finds that the packet must be transmitted to user network B. LER4 removes the label from the IP packet and forwards the packet to user network B.

Forward Equivalence Class (FEC): This is a group of IP packets that are forwarded in the same manner over the same path. Therefore, a group of packets with same FEC can be assigned the same labels through the path. Consider that multiple IP packets with the same forwarding characteristics are transmitted by network A to network B. LER1 assigns these IP packets the same labels throughout the path. These packets with same FEC will use the same path. This method reduces the size of the routing table.

MPLS defines only the forwarding methods so other protocols are needed to perform routing and signaling. The routing protocol distributes the networking

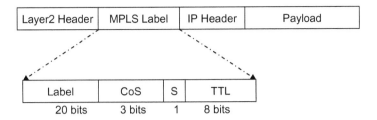

Fig. 10.19   MPLS label format

topology and configures the routing table by using an IP routing protocol such as OSPF, BGP, or RIP. Signaling protocols are used to inform the routers which label, and link, are to be used by the switch for each label switching path. MPL advantages are:

- The path from source to destination can be identified in advance.
- It can select network paths in order to have a balanced load in the network.
- MPLS can provide a specific path for data.
- MPLS offers quality of Service (QoS) by choosing a specific path in order to provide bandwidth to the application, less delay, and less packet loss.

MPLS label: The MPLS label is inserted in the frame but the location of the label depends on the layer 2 technology. In ATM, the label is VPI/VCI, and in Frame Relay, it is DLCI filed. In an IP packet, the label inserted between layer 2 and layer 3 is shown in Fig. 10.19. The MPLS label field is 32 bits and the following list describes the function of each field.

Label: This is 20 bits.
CoS (Class of Service): This field is 3 bits and is used for queuing and discarding packets traveling through the network.
S (Stack): This field is one bit and it is used for multiple MPLS labels.
TTL (Time-to-Live): This field is 8 bits and works similar to TTL field in IPv4.

10.8   IP Multicast

IP multicast is used for sending an IP packet to a group of receivers in a single transmission. The applications of IP multicast are:

A. Video/audio conferencing
B. Service advertisement
C. Stock distribution
D. IPTV
E. Transferring a bulk of data to multiple receivers
F. File distribution of operating system images in a university

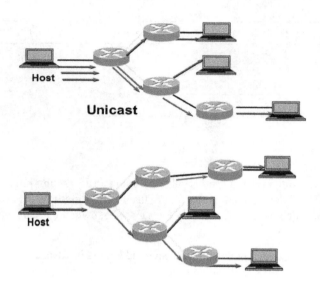

Fig. 10.20 Unicast and multicast diagrams

Figure 10.20 shows Unicast and Multicast. In Unicast, the source must send three packets, while in IP multicast, the source sends only one packet.

Characteristics of Multicast Network
- A multicast group is a set of hosts that all have the same IP multicast address.
- The transmitter of a multicast packet transmits the packet to a multicast router and the router transmits the packet to all hosts in the multicast group.
- A multicast router requires running Multicast Routing Protocol and Internet Group Message Protocol (IGMP).
- Each multicast group is identified by a single Class D IP multicast address.
- Members of a multicast group can be any place in a network.
- Each host in the network can join a multicast group.
- A multicast member (host) can leave the group by informing its multicast router.
- A receiver of a multicast packet must be a member of group.
- Any host on the network can transmit a multicast packet.
- Each multicast router manages the hosts connected directly to its port using Internet Group Managed Protocol (IGMP).

10.8.1   Internet Group Management Protocol—IGMP

The IETF published IGMP in RFC 1112. IGMP is a management protocol for IP multicast and is used for managing host group membership. It is located in the IP layer of the TCP/IP model as shown in Fig. 10.21. IGMP has three versions: IGMP v1, IGMPv2, and IGMPv3.

Fig. 10.21 IGMP

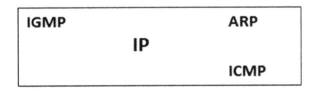

IGMP Messages
In general, IGMP has two types of messages:

E. Membership query message
F. Membership report message

A membership query message is transmitted by the router periodically to active, directly connected, hosts. This type of message is used for requesting information about host membership. Each active host member responds to the query with a membership report message. A Host sends an IGMP report to a router for joining a group. When a host wants to send a multicast packet to a group, it places its data in a UDP packet with a multicast IP address as the destination address. Figure 10.22 shows the IP multicast network.

A multicast router uses IGMP to learn which groups have members on each of their attached physical networks. It also keeps a list of multicast group memberships for each attached network.

IGMP Packet Format
Figure 10.23 shows the IGMP packet location in an IP packet, and Fig. 10.24 shows the IGMP packet format.

Type (8 bits): This field defines the type of packet as shown in the following table.

Type value in Hex	Description
0x11	Group membership query
0X12	IGMPv1 membership report
0x15	Cisco trace messages
0x16	IGMPv1 membership report
0x17	IGMPv2 leave group.

Max Response Time: This field is 8 bits and is used for membership. It defines the maximum allowed time before sending a responding report.

IGMP Checksum: This field is 16 bits and is used for error detection.

Group Address: This field is 32 bits. For a general query, this field set to zero such as in a membership query, and for a specific group, it is set to the group address.

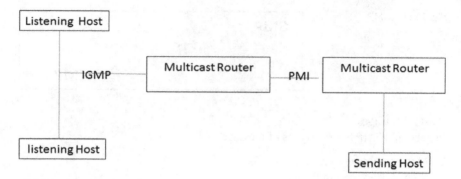

Fig. 10.22 IP multicast network

Fig. 10.23 IGMP packet

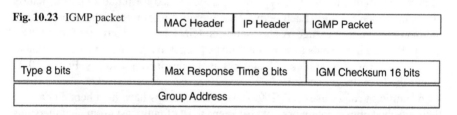

Fig. 10.24 IGMP packet format

10.9   Socket Programming

Socket programming is used for communications between a client and server through a network. A socket is a binding of an IP address and port number used for communications. The two processes communicate by sending their data into a socket, and then processing that data. Figure 10.25 shows the communications between a client and server using sockets.

The following steps show an implementation of socket programming for client and server using Python.

1. Import socket module
2. Create a socket
3. Binding the socket to an address and port
4. Connecting to a socket
5. Listening and accepting connections
6. Transferring and receiving data

Socket methods: The following are lists of methods used for socket programming:

s.recv()   used to receive data from a socket over TCP
s.send()   used to send data to a socket over TCP
s.recvfrom()   used to receive data from a socket over UDP

Fig. 10.25 Socket
programming between a
client and server

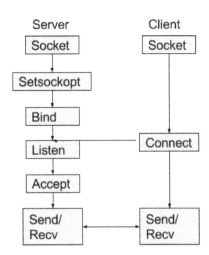

s.sendto()   used to send data to a socket over UDP

s.close()   closes a socket

s.gethostname()   gets the hostname of the machine that the program is currently
running on

1. Import the socket module

   import socket

2. Creating a socket

   A socket can be created by making call to the socket.socket() method.

s = socket.socket(socket.family, socket.type)

family: AF_UNIX is used for Unix, and AF_INET use for Internet

type: SOCK_STREAM is used for TCP protocol while SOCK_DGRAM is used for
UDP protocol

A. On the server side – create a TCP socket for listening

   s = socket.socket(socket.AF_INET, socket.SOCK_STREAM)

B. On the client side – create a socket to connect to

   s = socket.socket()

3. Binding the socket to an address and port. On the server side, the socket must be
bound to the host's IP address and port number by using bind(). If the server's IP
address is 192.168.1.2 and the desired listening port is 8000, then the binding
will be:

   s.bind(("192.168.1.2", 8000)

   In this model, the client does not require binding.

4. Connecting to the socket

Clients make a connection to the server by using **connect()** method.

s.connect(*address*)

where the address field is the IP address or hostname and port number of the host to connect to.

s.connect(("192.168.1.2", *8000*)).

5. Listening and accepting connections

The server must listen to its port for incoming connections by using the **listen()** method:

s.listen(5)

where the argument represents the number of connections, and therefore must be at least 1.

Once listening, the following function is used for accepting a connection on that socket:

clientsocket, address = socket.accept()

6. Transferring and receiving data

Data are sent and received by either the server and client using the send() or recv() methods:

s.send('Message12345'.encode())
data = s.recv(1024).decode()

Data must be encoded before being sent, and decoded at its destination as needed. The recv() method allows a buffer to be set, in this case receiving in chunks of 1024 bytes. With this process, the following outlines the creation of two python programs to be used to demonstrate socket programming, a client and a server capable of communicating across different computers on a local area network.

Computer 1:
1. Open command prompt.
2. Type "ipconfig".
3. Find your IPv4 address. Either write this down or keep the command prompt open.
4. We will now write code for the server using python. Start by opening notepad and typing the following code (everything after # is a comment and does not need to be typed):

> **import socket** # imports the socket module
> **serversocket   =   socket.socket(socket.AF_INET,   socket.SOCK_STREAM)**
> **host = "192.168.1.6"** # set this to the IPv4 address of Computer 1
> **port = 8000** # you can use any port as long as another service is not using it

```
serversocket.bind((host, port)) # binds the address and port to the socket
serversocket.listen(5) # has the socket start listening for incoming
connections
```

while 1:

```
(clientsocket, address) = serversocket.accept() # accept incoming
connection
data = clientsocket.recv(1024).decode()
print (data)
# send a message back to the client.
clientsocket.send("You have successfully connected".encode())
```

Now save this file on **Computer 1** as server.py and run it using Python. Keep this running, as the next step involves connecting to this server.

Computer 2:
Now we will create the client, open a new notepad, and type the following code:

```
import socket # imports socket module
s = socket.socket() # creates a new socket
host = "192.168.1.6" # set the host to the IPv4 address of Computer 1
port = 8000 # use the same port that was used on the server
s.connect((host,port)) # connects to the server
s.send('Message12345'.encode()) # sends a message to the server
data = '' # creates a variable to hold the incoming message
data = s.recv(1024).decode() # accepts and stores the incoming message in data
print (data) # prints the message
s.close # closes the connection
```

Save this file on **Computer 2** as client.py. Now run the client using Python. You should receive the message "*You have Successfully connected.*".

Summary

- DNS uses a tree hierarchy which consists of a root and TLDs.
- There are 13 DNS root servers.
- DNS clients use iterative queries or recursive queries to resolve a URL to an IP address.
- Each organization is required to have a Primary DNS and Secondary DNS server.
- DNS Resource Records contain a name, class, type and value.
- Dynamic Host Configuration Protocol (DHCP) is used to assign IP addresses to a host dynamically.
- The components of DHCP are: DHCP client, DHCP server, and DHCP relay agent.

- A DHCP server supports automatic allocation, dynamic allocation, and manual allocation for assigning IP addresses to clients.
- The application of DNS is to convert the URL or hostname of a server to an IP address.
- The function of a router is to determine the path for transmitted packets through the Internet.
- In a router, the path of information is determined by the routing Table.
- The routing table can be configured in two ways: static routing or dynamic routing.
- A static routing table is configured manually, and a dynamic routing table is configured automatically.
- MPLS uses label switching to route the information.
- A socket is a combination of IP address and port number.
- In socket programing, the client requests a connection.
- In socket programming, the server listens on its port for client requests.

Key Terms

Binding	Network control protocol
DHCP relay agent	Routing
DNS root servers	Routing information protocol
DNS root servers	Socket
Dynamic host configuration (DHCP)	Static routing
Dynamic routing	Top level domain
Multi-protocol label switching (MPLS)	Top level domain

Review Questions

Multiple Choice Questions

1. An organization requires:

    (a) One DNS server
    (b) Two DNS servers
    (c) Tree DNS servers
    (d) More than three DNS servers

2. How many DNS root servers exist?

    (a) 1
    (b) 10
    (c) 12
    (d) 13

3. Which of the following methods is used for resolving an IP address with a DNS server?

   (a) Iterative query
   (b) Recursive query
   (c) ARP
   (d) a & b

4. Which of the following protocol is used for DNS?

   (a) UDP
   (b) TCP
   (c) ARP
   (d) a & b

5. DHCP is used to assign _____ to a client.

   (a) IP addresses automatically
   (b) MAC addresses automatically
   (c) IP addresses manually
   (d) MAC addresses manually

6. A _____ holds a range of IP addresses.

   (a) client server
   (b) DHCP server
   (c) router
   (d) relay agent

7. The function of a relay agent is to_____.

   (a) pass the broadcast message
   (b) block the broadcast message
   (c) change the broadcast message to unicast
   (d) change the unicast message to multicast

8. DHCP Discovery packets are transmitted as_____ packets.

   (a) broadcast
   (b) unicast
   (c) multicast
   (d) a & b

9. DHCP Offer packets are transmitted as_____ packets.

   (a) broadcast
   (b) unicast
   (c) multicast
   (d) a & b

10. The function of a router is to_____.

   (a) find the destination IP address
   (b) find the destination MAC address
   (c) find the path to a destination for transporting packets
   (d) None of the above

11. Static routing tables are updated by_____.

   (a) a server
   (b) the network administrator
   (c) automatic method
   (d) none of the above

12. Dynamic routing tables are updated by_____.

   (a) a server
   (b) the network administrator
   (c) automatic methods
   (d) none of the above

13. MPLS uses _____ for routing of packets.

   (a) label
   (b) IP address
   (c) A & b
   (d) MAC address

14. A socket is a combination of a port number and.

   (a) IP address
   (b) Hostname
   (c) Default gateway
   (d) A & b

15. In a socket programming model, which device is listening to its port for incoming connections?

   (a) Server
   (b) Client
   (c) All station on network
   (d) Only server

Short Answer Questions

   1. What is the function of a DNS server?
   2. List the types of DNS queries.
   3. Show the DNS resource record format.
   4. What is the transport layer protocol used for DNS?
   5. What commands are used to clear the DNS cache table?
   6. What commands are used to display the DNS cache table?

7. List three diagnostic tools you can use for DNS troubleshooting.
8. List at least 5 TLDs.
9. What is the minimum number of DNS servers an organization must have?
10. Does an organization have separate DNS servers for IPv4 and IPv6?
11. What is the application of DHCP?
12. List the components of DHCP.
13. Explain DHCP dynamic IP allocation.
14. List DHCP packets and explain the function of each.
15. What is the function of a relay agent?
16. List the link characteristics.
17. Explain static routing and dynamic routing.
18. What is the function of Internet Control Message Protocol (ICMP).
19. List three applications IP multicast.
20. What is the application of a socket?
21. Which device requests connection to a socket?
22. Socket is combination of IP address and_____.

Chapter 11
Voice over Internet Protocols (Voice over IP)

Objectives

After completing this chapter, you should be able to:

- Discuss the applications of VoIP.
- Describe the factors that impact voice quality using VoIP.
- Explain the operation of VoIP.
- Discuss standards and protocols used for VoIP.
- List the components of SIP.
- Show the SIP protocol architecture.
- Describe the SIP connection setup between end users.
- Calculate the minimum bandwidth requirement for VoIP.

Introduction

Public telephone systems are based on circuit switching networks, which allow for real time communication between users. **Voice over IP (VoIP)** technology enables data networks such as the Internet, LANs, and WANs to be used for voice communication. Since VoIP reduces the cost of voice communication, it has become high in demand for corporations and organizations which have multiple locations. Corporations can avoid paying extra telephone charges by setting up a VoIP network for long-distance communication between office locations. Voice quality is an important factor in the success of Voice over data networks; it is imperative that VoIP offers the same quality as Voice over the **Public Switch Telephone Network (PSTN)**.

11.1   Voice Quality

There are three factors that impact voice quality over data networks: transmission delay, jitter, and packet loss.

Transmission Delay (Latency)
Transmission delay, or **latency**, is the time it takes a packet to travel one way from its destination to its source. For a one-way transmission, a transmission delay of between 150 and 250 m/sec is acceptable. This delay is generally caused by the following factors: propagation delay, the storage and forwarding of packets in routers and gateways, compression at the source, and decompression at the destination. When a voice packet is received late at the destination, it will just be discarded; the loss of a packet reduces the quality of the service. The latency can be reduced in private networks (Intranet, LAN) by adding quality of service (QoS) priority to the voice packets. However, the latency can not be controlled for the transmission of voice packets over the Internet.

Jitter Delay
Jitter delay is the difference in arrival time between packets. It is desired that the average arrival time between packets is constant. A variable delay is caused by congestion on networks and by voice packets being sent over different paths. If voice packets are received by the destination at irregular times, distortions in the sound will occur.

Packet Loss
Voice packets are transmitted over UDP, and therefore there is no guarantee that packets will reach their destinations. Packets may also be dropped by gateways when there is congestion in the network. When a packet is dropped, the gateway inserts a silence packet instead; this will result in gaps in the conversation. VoIP can tolerate about a 2% packet loss.

11.2   Applications of Voice over IP (VoIP)

VoIP can be implemented on the Internet, a LAN, or a WAN. Currently, many corporations offer long-distance calls using Voice over Internet technology at one-fifth of the price of using the PSTN. VoIP offers voice communication in two ways:

VoIP for an Organization
Currently, VoIP is used for telephone calls between offices in an organization by using the organization's pre-existing LAN.

International Call Provider
Some corporations offer international call services at a lower price for their customers. Figure 11.1 shows a block diagram of an International Call Provider. If User A

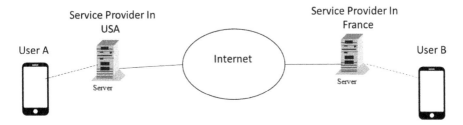

Fig. 11.1 Block diagram of an international call provider

would like to have a conversation with User B, then User A calls the server in the USA. Then, the server prompts User A for the destination phone number. User A enters 001 (international) and then 33 (France's country code), followed by the 10-digit phone number. The server in the USA then makes the connection to the server in France and transmits the 10-digit number to that server. The server in France finally dials the User B for communication.

11.3 Voice over IP Operation

Figure 11.2 shows the components and protocols used for transmitting voice packets over a data network. During this transmission, the following process will take place:

At the transmission side:

1. The microphone accepts the voice signal and passes it to the Pulse Code Modulation (PCM) section.
2. The PCM converts the voice signals to digital signals and passes the signals to the compression section.
3. The compression section compresses the voice bits and forms a voice packet. The voice packet is then passed to the **Real Time Protocol (RTP)**.
4. The RTP adds its header to the voice packet and passes the packet to UDP for transmission over IP.

At the receiving side:

1. The RTP passes its payload to the decompression section. The decompression section decompresses the voice packet and passes it to the analog-to-digital (A/D) converter.
2. The A/D converter converts the voice packet to analog and then passes it to the speaker.

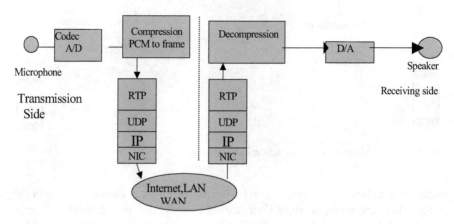

Fig. 11.2 Transmission of a voice packet

11.4   Voice over IP Protocols and Standards

Before a voice packet is transmitted over a data network, a connection between the two parties must exist. Currently, there are two protocols used for VoIP. They are H.323 developed by ITU and Session Initiation Protocol (SIP) developed by IETF. The functions of these protocols are to set up a connection, disconnect a connection, and handle call management. The SIP and H.323 protocols are used at the application level of the TCP/IP protocol.

Audio Codec
The function of an audio codec at the transmission side is to accept audio signals from a microphone and convert the audio signals to digital using an A/D converter. The audio codec then compresses, or encodes, the digital bits to form a voice packet. The function of the audio codec at the receiving side is to decompress, or decode, the voice packets and convert them to audio signals using a D/A converter and send the analog signal to the speaker.

Voice compression is performed by a device called a Vocoder (Voice Encoder/ Decoder), This device provides multiple types of voice compression. The type of compression is selected by a negotiation between the source and destination gateways. The following are some of the voice compression standards:

Standard	Data rate
G.723	5.3 kbps/6.3 kbps
G.729	8 kbps
G.711	Uncompressed 64 kbps

Real Time Protocol (RTP)
Real Time Protocol (RTP) is used for transporting audio and video packets over UDP. Figure 11.3 shows the RTP format.

2	1	1	4	1	7	16	bits
V	P	X	CC	M	PT	Sequence number	
Time Stamp							
SSRC							
CSRC							
Pay Load							

Fig. 11.3 RTP packet format

The following describes the function of each field in RTP packet format:

Version (V): Defines the RTP version.

Padding (P): When this field is set to 1, it means that extra bytes were padded to the payload. The last byte of the payload determines the number of the bytes that were padded the payload; these bytes should be discarded.

Extension (X): When this bit is set to 1, it means that that header is extended (for experimental use).

Contributing Source Count (CC): Used for multipoint call management.

Mark (M): This bit is to inform the receiver whether the packet is from a voice source or a video source. For voice applications, this bit is set for the first packet following silent suppression. For video applications, this bit is set only for the last packet of a video frame.

Payload Type (PT): Determines the type of payload.

Sequence Number: The receiver uses this number to correct any packets that were received out of order or to detect any packet losses.

Time Stamp: The time stamp depends on the payload. If the payload is a voice packet, then the time stamp is 8000, which is the sampling rate of the digitized human voice. If the payload is a video packet, then the time stamp is the clock rate for the video payload, which is 9000 Hz.

Synchronization Source Identifiers (SSRC): Used for multipoint calls.

Contributing Source Identifiers (CSSRC): Used for multipoint calls.

Real Time Control Protocol (RTCP): RTCP provides a control mechanism for jitter delay and packet loss in RTP; it is used for end-to-end monitoring of data delivery. The endpoints use RTP to exchange packets that carry voice data and periodically they exchange RTCP packets to monitor the quality of data exchange.

11.5   Session Initiation Protocol (SIP)

SIP is an Internet Engineering Task Force (IETF) Internet-based protocol designed for call set up and call management between two or more endpoints. SIP is a signaling protocol used for real time communication for VoIP such as Internet telephony and multimedia conferencing. SIP performs the following functions:

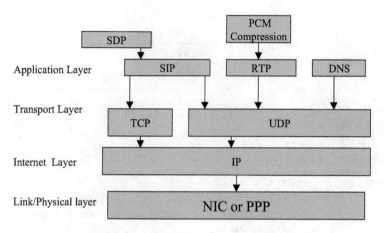

Fig. 11.4 SIP protocol architecture

1. Sets up calls between users.
2. Disconnects a call between callers.
3. Determines the location of the destination.
4. Supports addresses resolution (converting phone addresses to IP addresses).
5. Determines if the endpoint is available or not (busy).

   Figure 11.4 shows the SIP protocol architecture.

11.6   SIP Components

The components of SIP are user agents, gateways, and servers. The following describes the function of each component:

SIP User Agent or SIP Endpoint
The SIP user agent can be an IP phone or a PC with the SIP protocol. The user agent (UA) should be able to send a SIP request and response. The UA works in client and server mode. The UA also communicates with other user agents through a proxy server.

Gateway
A gateway is a special device that connects the PSTN to the Internet.

Server
SIP defines three types of servers: proxy servers, redirector servers, and registered servers. These three servers can be implemented in a SIP proxy server.

SIP Proxy Server   A SIP proxy server performs the following functions:

1. Accepts a user agent request and forwards the request to another user agent or server.
2. Accepts a response from a server or user agent and forwards it to a user agent.
3. If a proxy server does not have the IP address of the destination user agent, the proxy will contact a DNS server to obtain the IP address of the UA.
4. Requests a route on the behalf of user agent from a location server and also requests an IP address of the next proxy from a DNS server.

Proxy Server Types The SIP standard defines two types of SIP proxy servers: stateless proxy and stateful proxy:

A. **Stateless Proxy Server:** The stateless proxy server receives a request from a UA, processes the request, and forwards the response to a UA (user agent) or a server. The stateless proxy server does not keep any information (transactions) about the forwarded responses or requests. Therefore, if a response were lost due to congestion, the server would be unable to retransmit the response. The stateless proxy server is the simplest form of a proxy server.
B. **Stateful Proxy Server:** The server also acts as client when it responds to requests and sends requests. The stateful proxy server keeps information about responses and requests. Therefore, if a packet were lost due to congestion, the server would be able to retransmit the packet.

SIP Redirector Server The function of a redirector server is to accept requests and direct the client to contact alternate user agents (the same concept as when a secretary answers a phone and gives the caller another phone number). The redirector uses 3XX code to respond to a request. Some of the codes are as follows:

301 The destination has moved permanently.
302 The destination has moved temporarily, and the user is available at a different address.
305 The request source should contact the proxy server.

Registrar and Location Server The user agents register with a registrar server and the registrar server updates the location database (location server). The location database holds the address of the server that the UAs are connected to. Therefore, the proxy server can submit a client URL address to the location server and obtain an IP address of the user agent. The DNS server holds the IP addresses of the proxy servers.

11.7 SIP Request and Response Commands

SIP uses request and response commands to setup, change, or terminate a conversation between endpoints. A request is initiated by the client to the server, and a response is initiated by the server to the clients. The SIP entities use special words for requests and call methods. The following defines the methods and their descriptions:

Method	Description
INVITE	Used for inviting an endpoint for communication
BYE	Request for terminating a connection
ACK	Used for response to an invitation or for reliable communication between source and destination
REGISTER	Used by a user agent to register with a registrar server
CANCEL	Used for canceling a pending call
OPTIONS	Used for requesting information about call connection such as bandwidth or compression methods

Response Codes

SIP uses codes for responding to a request as well. The response codes are classified in as follows:

Class 1XX	Used to indicate progress such as ringing and searching
Class 2XX	Used to indicate success
Class 3XX	Used for redirecting and forwarding
Class 4XX	Client error
Class 5XX	Server failure
Class 6XX	Global failure (such as busy, not available)

The following table shows some of the specific response code and their descriptions:

Code	Description
100	Trying
180	Ringing
200	OK
301	Destination has moved permanently
302	Destination has moved temporarily
403	Not permitted
480	Unavailable
600	Busy
603	Declined

11.8   SIP Addressing

SIP uses addressing similar to e-mail addressing, such as by the usage of Universal Resource Locators (URLs) for addressing

sip: elahi@southern.edu
sip: +1-800-555-2020@xyz.com; user located in different network
sip: 25819@southernct.edu; user located in the same network

11.9   SIP Connection Operation

Consider a connection is made between two UAs through one proxy server. Figure 11.5 shows two UAs and a proxy server. SIP uses the following commands and responses in order to set up a connection between user A and B:

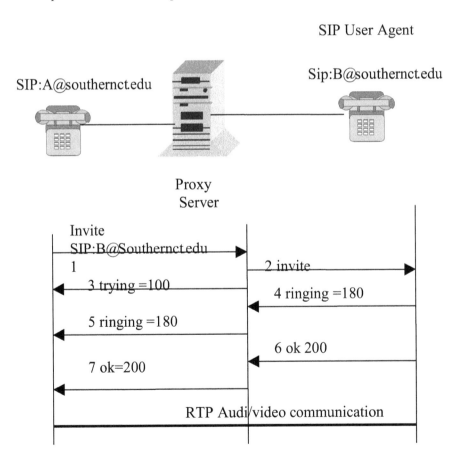

Fig. 11.5 Connections between two user agents and one proxy server

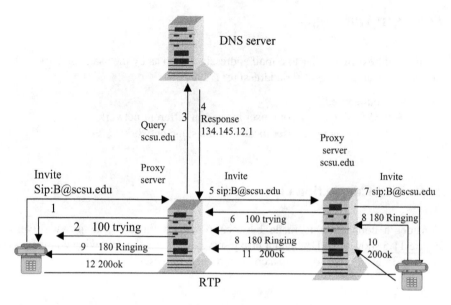

Fig. 11.6 Connection of two PSTN phones using the Internet

1. The user agent A sends a packet called an invite to the proxy server.
2. The proxy server accepts the packet and sends it to UA B.
3. The proxy server sends a code 100 to user A and user A waits.
4. Endpoint B accepts the invite packet, and it starts ringing (B agent is ringing).
5. The proxy server passes the code 180 to UA A and A generates rings indicating that user B is ringing.
6. When user B picks up the phone, B sends an ok packet with code 200 to the proxy server.
7. The proxy server sends an ok packet with code 200 to user A, user A stops ringing, and a session for communication using RTP is established between A and B.

Figure 11.6 shows a connection between two endpoints through two proxy servers. As shown in Fig. 11.6, the endpoint A invites endpoint B for connection through the proxy server A. The proxy server A does not have the IP address of endpoint B. Therefore, a query is sent to the DNS server to obtain the IP address of endpoint B.

11.10   Voice over IP Bandwidth Calculation

Voice packets must be received at a constant bit rate. The bit rate is dependent on the type of codec selection. The audio frame is made in 10 ms and the G.711 codec data rate is 64,000 bps. Therefore, in 10 ms, 640 bits or 80 bytes can be encoded.

IP header 20 bytes	UDP header 8 bytes	RTP header 12 byes	Voice payload Variable

Fig. 11.7 Voice payload

The G.792 data rate is 8000 bps, which means that 10 ms at this data rate is 80 bits or 10 bytes. The voice packet goes through RTP, UDP, and IP protocols, and these protocols all add their headers to the voice frame as shown in Fig. 11.7.

The voice packet goes through the Network Card (Ethernet, PPP, or Frame Relay) and the Network Card also adds its header and trailer, assuming an Ethernet NIC is used. Ethernet has a 22-byte header and a 4-byte trailer, which gives a total of 26 bytes added to the voice packet. Therefore, the total voice packet with a 10 ms payload using G.711 as a codec is calculated as follows:

Voice packet = 80 bytes payload + 40 Bytes (RTP, UDP and IP) header + 26 Ethernet header = 146 bytes

The voice packet must reach its destination at 64,000 bits per second or 8,000 bytes per second. The 8,000 bytes per second is equal to 100 voice packets per second. Therefore, the bandwidth of channel should not be less than:

Bandwidth of a channel =146 bytes*100 = 116800 bps

Summary

- VoIP reduces the cost of long-distance communication.
- Voice quality plays an important role in VoIP.
- Factors that impact voice quality are transmission delay, jitter, and packet loss.
- Voice is converted by a PCM to digital signal.
- Digital signals are converted to analog signals by an A/D converter.
- Protocols for VoIP are SIP and H.323
- Some examples of audio codecs are G.711, G.722 and G.723
- RTP is used for transporting audio and video packets over UDP.
- SIP is an IETF protocol for VoIP.
- SIP components are the SIP user agent, SIP gateway, and SIP proxy server.

Key Terms

Audio codec	Session Initiation Protocol (SIP)
Call signaling	SIP proxy server
H.323	SIP redirector server
Jitter Delay	SIP user agent
Multipoint Conferencing Unit (MCU)	Transmission delay
Real Time Control Protocol	Video Codec
Real Time Protocol (RTP)	VoIP

Review Questions

Multiple Choice Questions

1. Advantages of VoIP are reduced _____.

    (a) cost
    (b) delay
    (c) packet loss
    (d) none of the above

2. One of the factors that plays an important role in successful VoIP is:

    (a) Cost
    (b) Quality of service
    (c) Speed
    (d) Delay

3. The protocols currently used for VoIP are:

    (a) TCP and SIP
    (b) SIP and H.323
    (c) UDP and H.323
    (d) TCP and UDP

4. The Internet Engineering Task Force approved _____ for VoIP.

    (a) TCP
    (b) SIP
    (c) H.323
    (d) RTP

5. The function of an audio codec at the transmitting side is to _____.

    (a) convert digital signals to analog
    (b) convert voice to digital signal and compress
    (c) convert analog signals to digital
    (d) none of the above

6. Real Time Protocol is used for_____.

    (a) transporting data
    (b) transporting voice
    (c) transporting audio and video packet
    (d) transporting images

7. The SIP gateway is used to connect a _____ to the Internet.

    (a) LAN
    (b) WAN
    (c) PSTN
    (d) DSL

8. A _____ accepts a SIP user agent request and forwards it to another user agent.

(a) SIP endpoint
(b) SIP gateway
(c) SIP proxy server
(d) SIP redirector server

9. A _____ accepts requests and directs the client to contact the alternate user agent.

(a) SIP endpoint
(b) SIP gateway
(c) SIP proxy server
(d) SIP redirector server

Short Answer Questions

1. What does VoIP stand for?
2. Does VoIP reduce or increase costs of voice communications?
3. Why does VoIP reduce the cost of long-distance communications?
4. What is the most important factor to consider in VoIP?
5. What are three factors which impact voice quality over the Internet?
6. Define transmission delay and describe what causes it.
7. Define jitter and describe what causes it.
8. Define packet loss and describe what causes it.
9. List the applications for VoIP.
10. What does PCM stand for and what is the function of PCM?
11. Describe the layers involved for transmission of VoIP.
12. What are the two protocols used for VoIP?
13. What is the function of an audio codec at the transmission side?
14. What is a voice compression device called?
15. List the components of SIP.
16. What is the difference between a stateful and a stateless SIP proxy server?
17. What form of addressing does SIP use?
18. Describe the SIP connection operation.
19. Find the minimum bandwidth requirement for a VoIP channel using a 50 ms frame and a G.711 codec.
20. Find the minimum bandwidth requirement of a VoIP channel using a 20 ms audio frame and a G.729 codec.

Chapter 12
Wireless Local Area Network (WLAN)

Objectives

After completing this chapter, you should be able to:

- Discuss the applications and advantages of a wireless LAN (WLAN).
- Understand wireless LAN technology.
- Describe the applications of the ISM and UNII bands.
- Explain the operation of physical layers for a WLAN.
- Explain the access methods for WLANs.
- Distinguish between different types of IEEE 802.11.
- Discuss wireless LAN security.

Introduction

The **wireless local area network (WLAN),** or IEEE 802.11, is a LAN technology that enables users to access an organization's network from any location inside the organization without any physical connection to the organization's network. WLAN uses radio frequency or infrared waves as transmission media. The WLAN is the next generation of campus networks. Students are able to connect their laptops to the campus network from any location inside the campus. In hospitals, a WLAN allows doctors and nurses to access patients' files from any site in the hospital. Likewise, WLANs are used in warehouses and workshops. The following are some of the advantages of wireless LANs over wired LANs:

1. Wireless LANs can be used in places where wiring is impossible.
2. Wireless LANs can be expanded without any rewiring.
3. Wireless LANs provide the users mobility, that is, the users can move their computers anywhere inside the organization.
4. Wireless LANs support roaming allowing users to move around with their laptops without interrupting their connections.
5. Wireless LANs are cost effective as they make it possible to move from one location to another without the expense of connecting wires.

© The Author(s), under exclusive license to Springer Nature Switzerland AG 2024   243
A. Elahi, A. Cushman, *Computer Networks*,
https://doi.org/10.1007/978-3-031-42018-4_12

12.1 WLAN Components

Wireless LANs are composed of three main components: the WLAN Network Interface Card (NIC), Access Point, and Network Operating System.

A. **WLAN Network Interface Card (NIC):** The WLAN Interface Card operates at the data link layer. The MAC frame is transmitted to the physical layer, and then the physical layer changes the bits to a radio frequency (RF) signal for transmission. Also, the physical layer accepts RF signals converted to bits.

B. **Access Point (AP):** An Access Point is a wireless hub. It is connected to a wired LAN. The AP provides coordination between wireless devices. Figure 12.1 shows an AP. The antenna of the AP accepts a signal (acting as a receiver) and transmits signals (as a transmitter). The combination of the receiver and transmitter is called a transceiver.

Access Point (AP) operation

The Access Point operates in layer 2 of the OSI model, and its function is similar to a switch or a bridge:

1. A client requests association with an AP, then the AP authenticates the client and records the client's MAC address, informing the client that the association was accepted.

2. The client then transmits a packet through air to the AP, and the AP checks the destination MAC address of the frame. If the destination MAC address is in the AP's MAC table, then the AP transmits the frame through the air to its destination. Otherwise, if the destination MAC address is not in the MAC table, then the AP will change the frame format from IEEE 802.11 to an Ethernet II frame and transmit the frame to the connected wired LAN port.

Antenna An antenna is a conductor that it is used to radiate electromagnetic waves and to receive electromagnetic waves. The antenna can be characterized by directionality and gain. Directionality refers to the direction of the RF signal transmitted by the antenna. The two types are: the **omnidirectional antenna**, which transmits the RF signal through 360 degrees, and the **directional antenna**, which transmits in specific direction, as shown in Fig. 12.2.

Fig. 12.1 A wireless hub serves as the Access Point (Courtesy Dlink Corp)

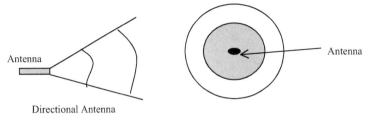

Directional Antenna

Omni-directional Antenna

Fig. 12.2 Directional and omnidirectional antenna

Antenna Gain   The Antenna Gain is measured in dBi, where dB stands for decibel and i stands for isotropic. An isotropic antenna is an ideal antenna that transmits the RF signal in all directions equally; but real antennas do not transmit RF signals in all directions. Therefore, the gain of antennas is given by Eq. 12.1

$$G = Pa / Pi \qquad (12.1)$$

where:

G is the Antenna Gain (measured in dBi)

$$G\,dBi = 10\log_{10} G$$

Pi: power density of an isotropic antenna at the same distance
Pa: power density of a real antenna in specific direction and distance

where:

Pi is defined by Eq. 12.2

$$Pi = Pt / 4pi\, r^2 \qquad (12.2)$$

C. Network Operating System: Most operating systems come with a wireless NOS.

12.2   WLAN Topologies

WLAN topologies can be classified as managed wireless networks and unmanaged wireless networks.

Managed Wireless Network Topologies   In a managed wireless network topology, the AP coordinates the communications between users. Topologies for managed wireless networks are the **Basic Service Set (BSS)** topology and the **Extended Service Set (ESS)** topology.

Basic Service Set (BSS)   BSS is composed of a group of workstations that can access each other and the wired LAN through an AP, as shown in Fig. 12.3. The

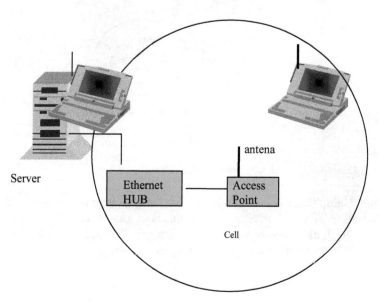

Fig. 12.3 Basic Service Set topology

function of an AP is to coordinate communication between Wireless LAN clients and the wired LAN. Also, the Access Point receives information from clients and retransmits that information to the hub. The AP works as a bridge between a wireless LAN and wired LAN. The area covered by an AP is called a **cell.** As shown in Fig. 12.4, any station inside the cell can access the AP. The Access Point coverage varies and depends on the manufacturer's design of the wireless LAN product, transmitter power, and the environment in which the WLAN operates (indoors or outdoors). When the number of clients in a cell or area of coverage increases, multiple Access Points will be added to the wireless LAN, which leads to an Extended Service Set (ESS) topology.

Extended Service Set (ESS) or Infrastructure Mode ESS is used to expand a wireless LAN. Figure 12.4 shows an ESS with multiple APs. The APs are all connected to a distribution system.

Distribution System (DS) The functions of the distribution system (DS) are to form extended service and to provide distribution services between Access Points. The DS basically acts like a bridge for Access Points.

AP Operation Modes APs can operate in the following modes:

- **Access Point**: This is the default mode of an AP, where wireless clients use an AP's Service Set Identifier (SSID) to transmit packets to the AP.
- **Wireless Repeater**: Will repeat the signal of another Access Point in order to extend the coverage.

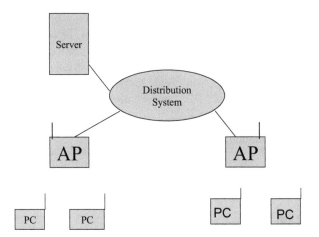

Fig. 12.4 Extended Service Set topology

Fig. 12.5 Wireless
Ad-Hoc topology

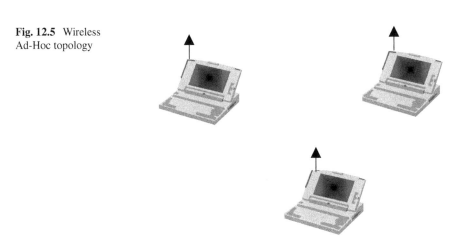

- **Wireless Bridge**: Will turn the Access Point into a wireless bridge. It will link a wireless network to a wired network.

Wireless Unmanaged Network   The topology for an unmanaged network is called **Ad-Hoc**. In an Ad-Hoc topology, the LAN is made of wireless devices without any Access Point. In this topology, each device communicates directly with other devices, as shown in Fig. 12.5.

12.3   Wireless LAN Technology

Two types of technology used for the transmission of information in WLANs are infrared (IR) technology and radio frequency (RF) technology.

Infrared Technology Infrared (IR) technology is suitable for indoor WLANs because infrared rays cannot penetrate walls, ceilings, or other obstacles. With infrared technology, the transmitter and receiver should see each other (be in line of sight) just like the remote control of a television set. In an environment where there are obstacles such as buildings, walls, etc., between the transmitter and a receiver, the transmitter may use diffused IR. However, most WLANs use RF technology.

Radio Frequency Technology There are two types of RF signals in use for the transmission of information: **narrowband signal** and **spread spectrum signal.**

Narrowband signal Narrowband signal refers to a signal with a narrow spectrum, as shown in Fig. 12.6. In narrowband, information is transmitted at a specific frequency, such as AM or FM radio waves.

Spread Spectrum In spread spectrum technology, information is transmitted over a range of frequencies, as shown in Fig. 12.7. Spread spectrum is one of the most popular technologies used for WLAN.

There are certain advantages of using the spread spectrum band over the narrowband. Some of these advantages are as follows:

- In spread spectrum technology, information is transmitted at different frequencies.
- It is hard to jam a spread spectrum signal; the signal cannot be disrupted by other signals.
- Interception of spread spectrum signals is more difficult than interception of a narrowband signal.
- Noise is less disruptive in spread spectrum signals than in a narrowband signal.

Fig. 12.6 Narrowband signal

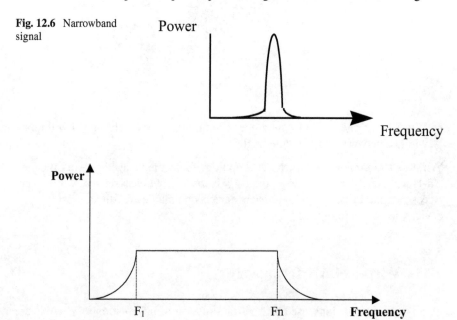

Fig. 12.7 Spread spectrum signal

12.4 WLAN Standards (IEEE 802.11 Family)

The IEEE 802.11 committee has approved several standards for WLAN. The standard defines functions of the Medium Access Control (MAC layer) and the Physical Layer. Table 12.1 shows the physical layer and data link layer for various WLAN standards.

12.5 Wireless LAN Physical Layer

In general, the physical layer performs the following functions:

(a) Modulation and encoding: Information is modulated and then transmitted to the destination.
(b) Supports multiple data rates.
(c) Senses the channel to see if it is clear or not (carrier sense).
(d) Transmits and receives information.

ISM Band The Federal Communication Commission (FCC) allocates a separate range of frequencies to radio stations, TV stations, telephone companies, and navigation and military agencies. The FCC also allocates a band of frequencies called the **industrial, scientific, and medical band (ISM)** for industrial, research, and medical applications. The use of the ISM band does not require a license from the FCC (with power of transmission up to one watt). Figure 12.8 shows the ISM band.

Table 12.1 IEEE 802.11 standards and their characteristics

MAC layer						
CSMA/CA and Point Coordination Function(PCF)						
IEEE802.11 physical layer						
Standard	Date	Frequency band GHz	Channel bandwidth	Modulation methods	Antenna	Data rate
802.11	1997	2.4	20 MHz	DSSS, FHSS		2 Mbps
802.11b	1999	2.4	20 MHz	DSSS		11 Mbps
802.11a	1999	5	20 MHz	OFDM		54 Mbps
802.11g	2003	2.4	20 MHz	DSSS, OFDM		54Mbps
802.11n	2009	2.4 and 5	20 ad 40 MHz	OFDM	Up to 4 MIMO	600Mbps
802.11ac	2013	5	20,40,80 MHz	OFDM	Up to 8 MIMO	6.93Gbps

902 MHz	928 MHz	2.4 GHz	2.48 GHz	5.725 GHz	5.85 GHz
Industrial Band I-band		Scientific Band S-band		Medical Band M-band	

Fig. 12.8 ISM band

12.5.1   IEEE 802.11 Physical Layer Operation

IEEE 802.11b and g operate at 2.4 GHz, while IEEE02.11n operates at both the 2.4 and 5 GHz band. The physical layer signal transmission methods are Frequency Hopping Spread Spectrum (FHSS), Direct Sequence Spread Spectrum (DSSS), Orthogonal Frequency Division Multiplexing (OFDM), and Orthogonal Frequency Division Multiple Access (OFDMA).

Frequency Hopping Spread Spectrum (FHSS)   The IEEE 802.11 standard recommends the use of the scientific band (2.4 GHz to 2.483 GHz) of the ISM band for WLAN. This band is divided into 79 channels of 1 MHz each. The transmitter sends each part of its information on a different channel. Figure 12.9 shows frequency hopping spread spectrum.

The order of the channels used by the transmitter to transmit information to the receiver is predefined, and the receiver knows the order of the incoming channels. For example, the transmitter may use a hop pattern of 3, 6, 5, 7, and 2 for transmitting information. The hop sequence can be selected during the installation of the WLAN. The FCC requires that a transmitter spend a maximum of 400 ms in each frequency for the transmission of data (this time is called the Dwell time) and use 75 hop patterns (each hop is one channel). The FCC also requires that the maximum power for the transmitter in the United States should not exceed one watt.

Frequency Hopping Spread Spectrum (FHSS) is more immune to noise because information is transmitted at different channels. In FHSS, if one channel is noisy, it can retransmit information on another channel.

Direct Sequence Spread Spectrum (DSSS)   In **DSSS,** before transmission, each bit of information is broken down to a pattern of bits called a **Chip.** For the generation chip bits, each information bit is Exclusive-ORed with **Pseudo Random Code,** as shown in Fig. 12.10. The output of the Exclusive-OR for each data bit is called chip bits. These chip bits are modulated and then transmitted. This method creates a higher modulation rate because the transmitter transmits the chip bit over a larger frequency spectrum. Figure 12.11 shows the transmission section of the physical layer. The receiver uses the same pseudo random code to decode the original data.

Power

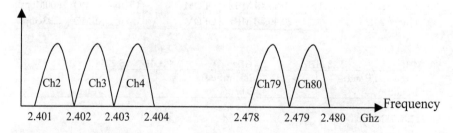

Fig. 12.9 Frequency Hopping Spread Spectrum

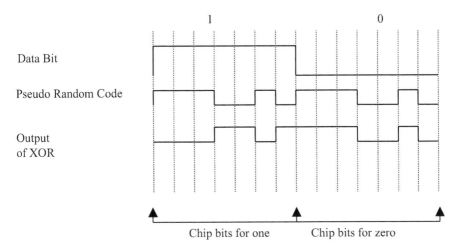

Fig. 12.10 Generation chips

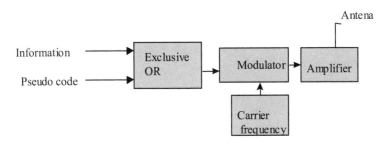

Fig. 12.11 Transmission section of physical layer

A larger chip sequence generates a larger frequency band. IEEE 802.11 recommends 11 bits for each chip.

The DSSS supports two types of modulation: Differential Binary Phase Shift Keying (DBPSK), which is used for a data rate of 1 Mbps, and Differential Quadrature Phase Shift Keying (DQPSK), which is used for data rates of 2 Mbps.

Orthogonal Frequency Division Multiplexing (OFDM) Multipath fading occurs when a communication signal transmitted through the air takes multiple paths to its destination. Due to this problem, multiple copies of the same signal will reach the intended target at different times. The problem may become compounded if this delay is greater than the time it takes to transmit a signal. This may cause what is known as Inter-Signal Interference (ISS), whereby a subsequent signal transmission arrives before a primary transmission. Applying OFDM to the transmitted signal will eliminate the problems associated with ISS.

OFDM, also called Multicarrier Modulation, divides the channel bandwidth into multiple orthogonal (out of phase by 90 degrees) subchannels, as shown in Fig. 12.12. The data stream is divided into n*m bit streams, where n is the number

Fig. 12.12 OFDM
Subchannels

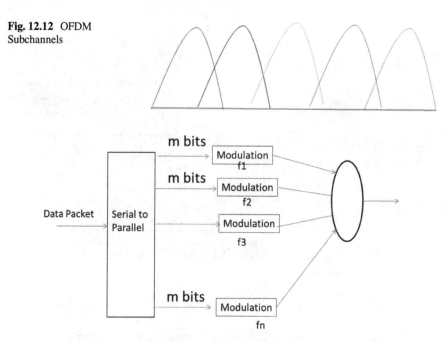

Fig. 12.13 Block diagram of OFDM

of channels and m is the number of bits in the data stream. The value of m depends on the modulation technique used by the Modulation block shown in Fig. 12.13. If QPSK is used for modulation, then m is equal to 2 bits. Each bit stream will use a different subcarrier for transmission.

OFDM has been used in IEEE 802.11a/g, IEEE 802.11n, IEEE 802.16a, and Digital Audio Broadcast (DAB).

Orthogonal Frequency Division Multiple Access (OFDMA) In OFDM, only one user transmits on a subcarrier channel. Multiple users transmit on subcarrier channels at different times, where each user is assigned subchannels for transmission.

12.5.2 IEEE 802.11b Physical Layer

The IEEE 802.11b standard extends the DSSS physical layer of 802.11 to provide higher data rates of 5.5 Mbps and 11 Mbps. 802.11b uses Complementary Code Keying (CCK that may be used) to support the two data rates: 5.5 Mbps and 11 Mbps, in addition to 1 Mbps and 2 Mbps.

IEEE 802.11b Channels The IEEE 802.11b standard defines 11 channels; each channel is represented by its center frequency. For example, channel 1's center frequency is 2412Mhz, and channel 2's is 2417 Mhz. The center frequency of each channel is separated from adjacent channels by 5 MHz. The bandwidth of each

AP(1)	AP(6)	AP(11)
AP(11)	AP(1)	AP(6)
AP(6)	AP(11)	AP(1)

Fig. 12.14   Locations of Access Points in a three-story building

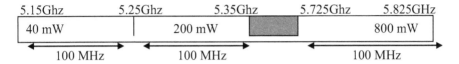

Fig. 12.15   Frequency of operation for U-NII band

Table 12.2   Characteristics of IEEE 802.11 a, b, and g

	802.11b	802.11g	802.11a
Available RF channel	3 nonoverlapping	3 nonoverlapping	8 nonoverlapping
Frequency band	2.4 GHz	2,4 GHz	5 GHz
Maximum data rate	11 Mbps	54 Mbps	54 Mbps
Rate/typical range	100ft at 11Mbps 300ft at 1Mbps	50ft at 54 Mbps 150 ft at 11 Mbps	40 ft at 54 Mbps 300 ft at 6Mbps

channel is 16 MHz, and using adjacent channels will cause interference. IEEE 802.11b supports three nonoverlapping channels, 1, 6, and 11, to overcome interference. To avoid channel overlap, the most common used channels are 1, 6, and 11. Figure 12.14 shows an arrangement of APs in a three-story building, where each row represents one floor with its Access Point and channel number.

IEEE 802.11a Physical Layer   IEEE 802.11a operates at 5 GHz and it is not compatible with IEEE802.11b; it uses DSSS technology. IEEE 802.11a operates in the **Unlicensed National Information Infrastructure Band (U-NII).** The U-NII band consists of three 100 MHz frequency bands, as shown in Fig. 12.15. The physical layer of IEEE 802.11a uses Orthogonal Frequency Division Multiplexing (OFDM) for transmitting data at higher rates. IEEE 802.11a offers data rates of 6, 9, 12, 18, 24, 36, 48, and 58 Mbps. The physical layer can use any of the BPSK, QPSK, 16QAM, and 64QAM for modulation depending on the data rate. The frequency of the operation is made of twelve 20Mhz channels.

IEEE 802.11g Physical Layer   IEEE 802.11a and IEEE 802.11b define different standards, which are not compatible with each other. IEEE 802.11b operates at 2.4 GHz and transmits data at the rate of 11 Mbps using DSSS technology, whereas IEEE 802.11a uses OFDM. IEEE 802.11g operates in 2.4 GHz using DSSS and OFDM for the transmission of information. Table 12.2 shows the characteristic of IEEE 802.11 a, b, and g.

12.6   Physical Layer Architecture

Figure 12.16 shows the physical layer architecture of the IEEE 802.11 family. The physical layer is divided into two sublayers: the Physical Layer Convergence Procedure (PLCP) and Physical Medium Dependent.

The MAC layer transfers a frame to the Physical Layer Convergence Procedure (PLCP). The PLCP adds its own header to the MAC frame and transmits the frame to PMD for transmission. The IEEE 802.11 defines frequency hopping and DSS for the physical layer. Figure 12.17 shows the PLCP header for DSSS, and Fig. 12.18 shows the PLCP frame format for frequency hopping spread spectrum (FHSS).

The following describes the function of each field in the PLCP frame format:

Sync Field: Sync Field is used for synchronization. It is 80 bits of alternating 0's and 1's.
SFD Field: The SFD Field is 00001100 10111101.
Length Field: This field defines the length of PLCP in bytes.
Signaling: This field indicates to the physical layer the modulation type that must be used for transmission of the frame. The data rate is calculated as follows:

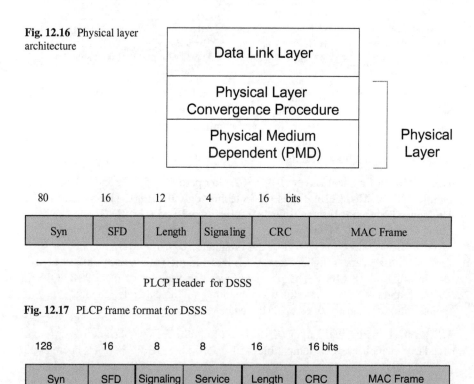

Fig. 12.16 Physical layer architecture

PLCP Header for DSSS

Fig. 12.17 PLCP frame format for DSSS

Fig. 12.18 PLCP for FHSS

	Value of signal field in Hex	Data rate in Mbps
Table 12.3 Value of signaling field and data rate	0A	1
	14	2
	37	5.5
	6E	11

$$\text{Data rate} = \text{Value in signal field}^* \, 100 \, \text{Kbps}$$

The value of signaling field and its data rate is shown in Table 12.3.

CRC-16: Is used for error detection in the PLCP header.
Service Field: This field is reserved.

IEEE 802.11b has two types of PLCP Preamble headers, Short PLCP preamble headers and Long PLCP preamble headers. The short PLCP Preamble header has 56 bits for the Synchronization field and the Long Preamble header has a 128-bit Synchronization header.

Interframe Space
In general, the Interframe space enables the receiver to complete the frame before the next frame comes. The IEEE 802.11 defines three types of Interframe Spaces (IFS) among the frames transmitted between source and destination. They are:

1. Short Interframe Space (SIFS): This Interframe Space is used for immediate responses such as ACK, CTS and RTS.
2. Distribution Coordination Function Interframe Gap Space (DIFS): DIFS is used for the spacing of data frames.
3. Point Coordination Function Interframe Space (PIFS): This interval is used for the point condition access method and the gap is used for polling of a client. The client should respond after this time.

12.7   WLAN Medium Access Control

The Medium Access Control (MAC) layer performs the following functions:

- Supports multiple physical layers
- Supports access control
- Fragmentation of frame
- Frame encryption
- Roaming

IEEE 802.11a defines distribution coordination function (Carrier Sense Multiple Access with Collision Avoidance CSMA/CA) and point coordination function as methods for a station to access Wireless LANs.

Fig. 12.19 CSMA/CA
flowchart

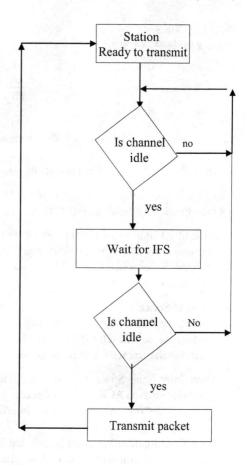

Carrier Sense Multiple Access with Collision Avoidance

Carrier Sense Multiple Access with Collision Avoidance (CSMA/CA) is similar to
CSMA/CD. In CSMA/CA, when a station wants to transmit a frame, first it listens
to the medium. If there is no traffic, it continues to wait for a **Short Interframe
Space,** and if there is still no traffic on medium, then the station will start transmitting. Otherwise, it has to wait for the medium to become clear. Figure 12.19 shows
the CSMA/CA flowchart operation.

In order to reduce the probability of two stations transmitting simultaneously
because they cannot hear each other, the standard defines a **Virtual Carrier Sense**
mechanism.

Virtual Carrier Sense Figure 12.20 shows two stations and one Access Point.
Stations B and C are covered by an Access Point, but station B cannot cover station
C. While the Access Point communicates with station B, C can listen to the AP and
detect if that medium is clear or not. Station C waits for the media to be clear. When
B is transmitting to the AP, C is not in the range to detect that B is transmitting to
the AP. Therefore, C will see the media as clear and start transmission. B and C will

Fig. 12.20 CSMA/CA
for WLAN

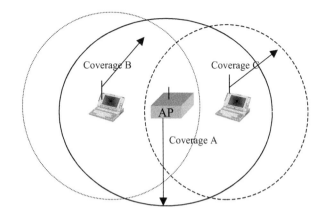

Fig. 12.21 CSMA/CA
process

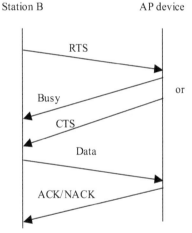

be transmitting to the AP at the same time and cause a collision. Station C is a hidden station; therefore, there is no physical connection to detect this collision. The following steps describe the CSMA/CA operation and Fig. 12.21 shows the CSMA/CA process.

1. Station B wants to transmit to the AP so it senses the medium. If the medium is clear, it sends a short message to the AP called the Request to Send (RST). This message contains the destination and source addresses, and the size of the data to be transmitted.
2. If the AP is ready to communicate with B, the AP will send a Clear to Send (CTS) frame to B; otherwise, it will send a busy frame. This signal can be detected by station C and is taken as a busy medium. Station B receives a CTS signal and then transmits its frame. The receiver acknowledges each frame transmitted by B.

Point Coordination Function (PCF) Access Method

IEEE 802.11 provides the point coordination function as an optional access method. The AP polls each user for transmission. In the PCF method, the AP listens to the medium; if there is no traffic in the medium it sends a **Beacon** frame to all users indicating that the PCF method will be used for the access method and all users must stop transmission. Then the AP sends a poll frame to a specific user. If the user has any data, it will transmit; otherwise, the user will send a null frame to the AP.

12.8   MAC Frame Format

IEEE 802.11 defines three types of frames for WLAN: Management frames, Control frames, and Data frames. Figure 12.22 shows the MAC frame format for IEEE 802.11.

The function of each field in MAC frame of IEEE 802.11 are described as follows:

Frame Control: The Frame Control field is 2 bytes and defines the type of frame, as shown in Fig. 12.23.

The following describes the function of each field of Fig. 12.23:

Protocol: Defines the protocol version. The current version is zero.

Type: Defines the type of the frame. 00 means management frame, 01 means control frame, and 10 means data frame.

Subtype: Defines the subframe in each type. There are several subtype management frames.

To DS and From DS Fields: To DS and From DS define the direction of the frame and the function of the address fields in Fig. 12.23. Table 12.4 describes the function of To DS, From DS, and address fields (Address1, Address2, Address 3, and Address 4 fields).

In Table 12.4, DA is the destination address (MAC address) and SA is the source address.

More Fragment: The fragment field set to 1 indicates that more frames, belonging to the same application, are coming. This field set to zero indicates to the destination that the current frame is the last frame.

Power Management: This bit set to 1 indicates that the transmitter is operating under power management.

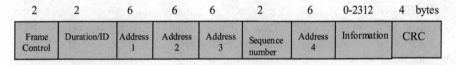

Fig. 12.22 MAC frame for IEEE 802.11

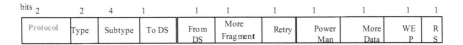

Fig. 12.23 Frame Control field

Table 12.4 To DS and from DS

To DS	From DS	Address 1	Address 2	Address 3	Address 4	Comments
0	0	DA	SA	BSS ID	Unused	Transferring a frame between two stations located in the same BSS or Cell
0	1	DA	BSS ID	SA	Unused	Transferring frame from DS to a station in a cell
1	0	BSS ID	SA	DA	Unused	Transferring a frame from a station to DS
1	1	Receiver Address	Transmitter Address	DA	SA	This is used for wireless distribution system

Wired Equivalent Privacy (WEP): This bit set to 1 to indicate that a Cryptographic Algorithm has changed the information.

Retry: Retry set to 1 indicates that this is a copy of the previous frame.

RS: Reserved bit.

Duration/ID   This field contains the duration value (Network Allocation Vector NAV value) to inform other stations how long it will take for the source to complete its transmission (in microseconds). The other stations use the NAV value to defer their transmissions.

Sequence Number   This field is divided into two fields: a 4-bit field and a 12-bit field. The first four bits indicates a fragment number, and the 12-bit field indicates the sequence number of the frame.

12.9   WLAN Frame Types

The IEEE defines three types of frames: management frames, control frames, and data frames.

Management Frames   The management frame is used by a station to make a connection to the AP, to disconnect the station from the AP, and for timing and synchronization. Some of the management frames are:

- **Association Request** (subtype = 0000): Client sends a request frame for joining a BSS network.
- **Association Response** (subtype = 0001): AP responds to client request as to whether the AP is accepting this request or not.

- **Reassociation** (Subtype = 0010) Sending a frame by a client when moving from one BSS to another.
- **Reassociation Response** (Subtype = 0011): In response to reassociation, AP sends a reassociation frame to the client on whether it accepts the client joining the AP to BSS or not.
- **Disassociation** (Subtype = 1010): Used by a client to terminate its association with BSS.
- **Beacon** (Subtype = 1000): A Beacon frame is transmitted by the Access Point periodically to inform other wireless stations of its presence. This frame contains the following information:

  (a) The Beacon interval, which is used by the station to know when to wake up for the next beacon
  (b) Timestamp, which is used for synchronization between the Access Point and wireless station
  (c) Types of signaling, such as FHSS and DSSS
  (d) Supported data rates

- **Probe Request** (Subtype = 0101): Station requesting information from another station.

Control Frames Used for flow control such as positive acknowledgement (ACK), RTS (Subtype = 1011), and CTS (Subtype = 1100).

Power Management Most of the Clients in a Wireless LAN are laptop computers and mobile phones, and power saving is an important factor for a wireless client. Therefore, IEEE 802.11 defines two power management modes. They are: Active or Continuous Aware Mode (CAM) and Power Save polling (PSP) "sleep" mode.

Active Mode or Continuous Aware mode (CAM): In this mode, the wireless Client uses full power.

Power Save polling (sleep): In this mode, the wireless client goes to sleep, meaning that the client uses less power and turns off power to the display, disks, and other peripherals that are not needed.

In PSP mode, the following operations take place:

(a) The client sends a frame to the AP informing the AP that the client is going to sleep.
(b) The Access Point records the stations that are asleep.
(c) The Access Point buffers the sleeping client's frames.
(d) The clients that are asleep continuously receive beacon frames.
(e) The Access Point sends beacon frames with traffic indication map (TIM) which informs clients that were asleep that they have frames in the AP.
(f) The clients that have frames in the AP switch their power mode to active mode.
(g) The client sends a request for its frames.

Data Frames Data frames are used for the transfer of information.

12.10   Station Joining A Basic Service Set

In order for a client to associate with an Access Point, it must perform the probe phase, authentication phase, and association phase. The following describes each phase in detail:

Probe phase: When a station wants to join a BSS, the station needs to get synchronization information from the AP (clock value of AP). This can be accomplished in two ways: by passive scanning and active scanning.

> *Passive scanning*: In this method, the station can receive a beacon frame, which is being sent out periodically by the Access Point. This beacon frame contains synchronization information.
>
> *Active scanning:* In this method, the station transmits a Probe request frame to locate an Access Point and waits for a Probe response. The Probe response frame contains the synchronization clock.

Association phase: If the authentication phase is completed successfully, the station will send an association request packet to the Access Point. The Access Point adds the station to its association table. A station can associate only with one Access Point at a time.

12.11   Roaming

Roaming is when a station moves from one cell to another cell without losing connection. In WLAN, moving from one cell to another must be performed between packet transmissions, meaning the packet must be transmitted completely before moving to another cell.

12.12   Wi-Fi Certification

Since there are many manufacturers of wireless equipment, it is necessary to ensure that all wireless devices are interoperable. Therefore, the Wireless Ethernet Compatibility Alliance (WECA) offers certification for wireless equipment interoperability, referred to as **Wireless Fidelity (Wi-Fi).**

12.13   WLAN Signal Distortion

The RF signal can be distorted while going through physical obstacles such as a wall, a ceiling, or by multipath fading.

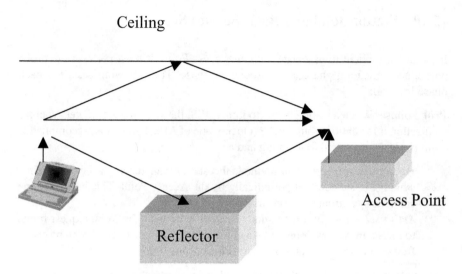

Fig. 12.24 Multipath fading

Multipath Fading Multipath fading occurs when a communication signal transmitted through the air takes multiple paths to reach the destination, as shown in
Fig. 12.24. The receiver receives multiple signals with different delays and amplitudes. When the delay of the transmitted signal increases, it will distort the signal
causing data to be corrupted and reduce data throughput.

12.14   IEEE 802.11n

The IEEE 802.11n protocol adds several enhancements to the physical and MAC
layers of the IEEE802.11a/b/g protocols in order to improve performance, efficiency, and throughput. These enhancements are:

1. Multiple-Input Multiple-Output (MIMO): MIMO uses multiple transmitters and
   multiple receivers. MIMO, as defined under IEEE 802.11n, is characterized by
   the formula N*M, where N is the number of transmitters and M is the number of
   receivers ranging from 1*1 to 4*4.
2. 40 MHZ operation band.
3. Frame aggregation: Combining two or more frames into a single frame for
   transmission.
4. Block acknowledgment: Multiple packets can be acknowledged with single packet.
5. Backward compatibility with IEEE 802.11 a/b/g.
6. Spatial multiplexing: Simultaneously send multiple data stream to the receivers.

7. Beamforming transmissions: Used to improve the signal to noise (S/N) ratio in order to increase the coverage area. This is applicable only for the use of one receiver (IEEE80.11a/b/g).

The following are IEEE 802.11n characteristics:

A. Ratified date: 2009
B. Operation band: 2.4 GHz and 5 GHz
C. Modulation: up to 64 QAM
D. Channel bandwidth: 20 and 40 MHz
E. MIMO: up to 4*4
F. Maximum data rate: 600 Mbps
G. Technology: OFDM
H. 24 channels (20 MHz channels)
 I. 12 channels (40 MHz channels)

MIMO (Multiple-Input and Multiple-Output)
The most important development of IEEE 802.11n is the use of MIMO (Multiple-Input and Multiple-Output) technology. MIMO technology consists of beamforming and Spatial Multiplexing.

Beamforming
According Nyquist's theorem, the maximum data rate of a communication channel is defined by the equation:

$$MAX \text{ Data Rate} = W \log(1 + S/N), \tag{12.3}$$

where S/N is the ratio of signal power to noise power.

The larger the S/N, the higher the channel data rate. MIMO technology uses beamforming to increase the S/N ratio. If the transmitted signals are of the same phase, the effect will be an increase of the amplitude of the signal at the receiver, as seen in Fig. 12.25. If the transmitted signals are out of phase, the effect will be a decrease in the amplitude of the signal seen at the receiver, as is seen in Fig. 12.26.

Beamforming is useful when there is only one receiver, and it is used to expand the coverage area of the signal.

12.15.20 MHz bands and 40 MHz Bands The IEEE 802.11a/g uses 20 MHz channel spacing with Orthogonal Frequency Division Multiplexing (OFDM). IEEE 802.11a/g divides the 20 MHz channel into 52 subchannels and uses 48 subchannels for transmission. IEEE802.11n combines two 20 MHz channels resulting in a single 40 MHz bandwidth channel. The 40 MHz channel is then divided into 114 subchannels, with 108 channels used for transmission, resulting in a doubling of the data rate.

Spatial Multiplexing
Spatial multiplexing subdivides a data stream into multiple pieces called Spatial Streams (SP). Each spatial stream is transmitted through a different transmitter using the same channel frequency. Figure 12.27 shows a 2*2 MIMO with a Data Stream divided into four spatial streams; M1, M2, M3, and M4. The M1 and M2

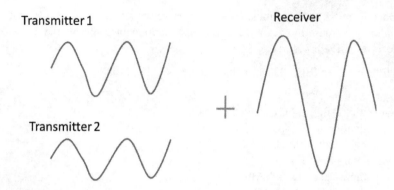

Fig. 12.25 Beamforming (same phase)

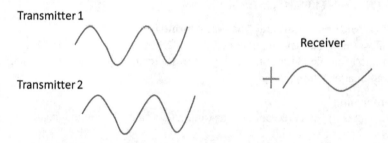

Fig. 12.26 Beamforming (out of phase)

Fig. 12.27 2*2 MIMO

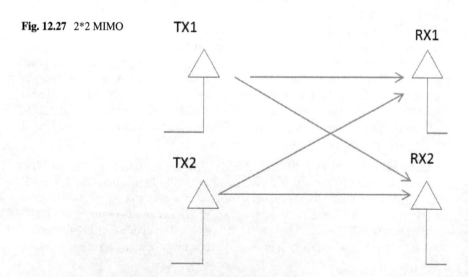

Spatial streams are transmitted by Tx1 and TX2, respectively, and the M3 and M4 spatial streams are transmitted by TX1 and TX2, respectively. The receiving side reassembles the streams.

Modulation and Coding

The IEEE 802.11n protocol will negotiate its capability with clients for selection of spatial stream, RF modulation, coding rate, and guard interval (guard interval is the time between transmitted frames). The guard interval of the IEEE 802.11a/g protocol is 800 ns, but the guard interval of the IEEE 802.11n protocol is 400ns.

Frame Aggregation

IEEE 802.11n combines multiple frames into a single frame for transmission. There are two types of frame aggregation:

1. Aggregated MAC Service Data Unit (A-MSDU)
2. Aggregated MAC Protocol Data Unit (A-MPDU)

A-MSDU: A-MSDU is used by a station that has multiple data packets to send to a single destination. The station combines the data packets into a single MAC Protocol unit for transmission as shown in Fig. 12.28. In Fig. 12.29, all three packets are combined by the MAC layer, which then adds its own header.

A-MPDU: A-MPDU is used by a station that has multiple data packets to send to a single destination but which originates from different applications. Each packet has its own MAC header but the combined packet uses only one physical layer header as shown in Fig. 12.29.

Block Acknowledgment

Block acknowledgment is used to acknowledge multiple frames. If the receiver receives three frames, then it acknowledges all three frames with one frame, as shown in Fig. 12.30.

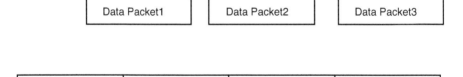

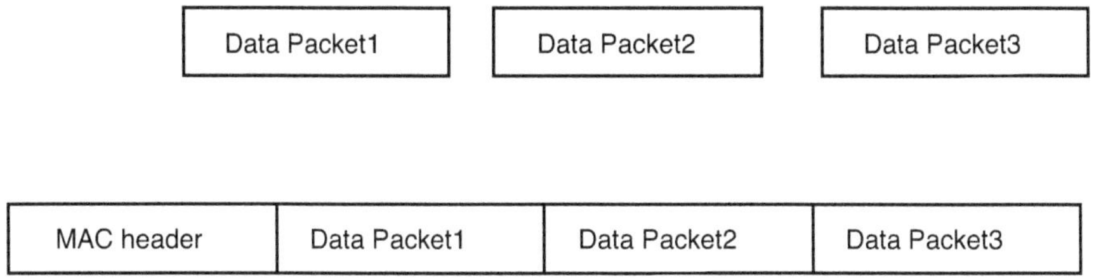

| Data Packet1 | Data Packet2 | Data Packet3 |

| MAC header | Data Packet1 | Data Packet2 | Data Packet3 |

Fig. 12.28 A-MSDU aggregation

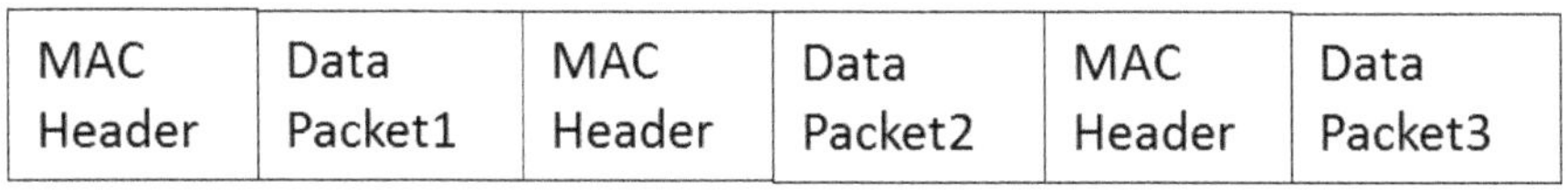

| MAC Header | Data Packet1 | MAC Header | Data Packet2 | MAC Header | Data Packet3 |

Fig. 12.29 A-MPDU aggregation

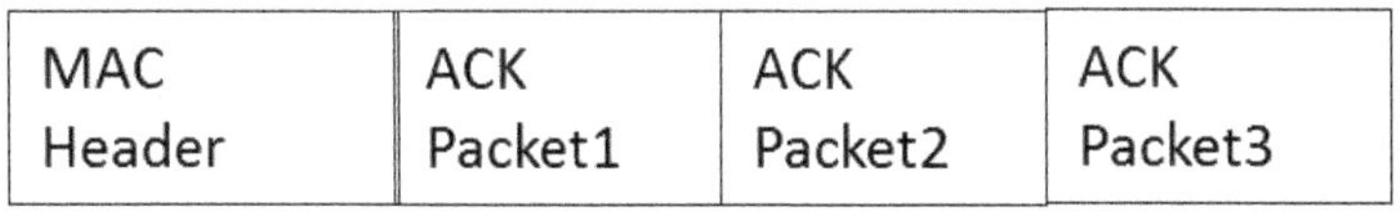

| MAC Header | ACK Packet1 | ACK Packet2 | ACK Packet3 |

Fig. 12.30 Block acknowledgment

Guard Interval (GI)

The guard interval is the space between sequentially transmitted frames. It is used to ensure that frames taking a longer path will not collide with newly transmitted frames. IEEE 802.11a/g defines an 800 ns guard interval but IEEE 802.11n uses a 400 ns guard interval.

Coding Rate

The coding rate defines the useful data in a packet. Extra bits are added to packets for forward error correction. If the useful information is represented by K bits and N represents the total bits in the packet, then the code rate is denoted by K/N.

Data rates

Table 12.5 shows the data rate of IEEE 802.11n for different types of modulation, code rate, and GI.

Table 12.5 Data rate of IEEE 802.11n (wikipedia.org/wiki/IEEE_802.11n-2009)

MCSindex	Spatialstreams	Modulationtype	Codingrate	Data rate (Mbit/s)			
				20 MHz channel		40 MHz channel	
				800 ns GI	400 ns GI	800 ns GI	400 ns GI
0	1	BPSK	1/2	6.50	7.20	13.50	15.00
1	1	QPSK	1/2	13.00	14.40	27.00	30.00
2	1	QPSK	3/4	19.50	21.70	40.50	45.00
3	1	16-QAM	1/2	26.00	28.90	54.00	60.00
4	1	16-QAM	3/4	39.00	43.30	81.00	90.00
5	1	64-QAM	2/3	52.00	57.80	108.00	120.00
6	1	64-QAM	3/4	58.50	65.00	121.50	135.00
7	1	64-QAM	5/6	65.00	72.20	135.00	150.00
8	2	BPSK	1/2	13.00	14.40	27.00	30.00
9	2	QPSK	1/2	26.00	28.90	54.00	60.00
10	2	QPSK	3/4	39.00	43.30	81.00	90.00
11	2	16-QAM	1/2	52.00	57.80	108.00	120.00
12	2	16-QAM	3/4	78.00	86.70	162.00	180.00
13	2	64-QAM	2/3	104.00	115.60	216.00	240.00
14	2	64-QAM	3/4	117.00	130.00	243.00	270.00
15	2	64-QAM	5/6	130.00	144.40	270.00	300.00
16	3	BPSK	1/2	19.50	21.70	40.50	45.00
17	3	QPSK	1/2	39.00	43.30	81.00	90.00
18	3	QPSK	3/4	58.50	65.00	121.50	135.00
19	3	16-QAM	1/2	78.00	86.70	162.00	180.00
20	3	16-QAM	3/4	117.00	130.70	243.00	270.00
21	3	64-QAM	2/3	156.00	173.30	324.00	360.00
22	3	64-QAM	3/4	175.50	195.00	364.50	405.00
23	3	64-QAM	5/6	195.00	216.70	405.00	450.00
24	4	BPSK	1/2	26.00	28.80	54.00	60.00
25	4	QPSK	1/2	52.00	57.60	108.00	120.00

(continued)

Table 12.5 (continued)

MCSindex	Spatialstreams	Modulationtype	Codingrate	Data rate (Mbit/s)			
				20 MHz channel		40 MHz channel	
				800 ns GI	400 ns GI	800 ns GI	400 ns GI
26	4	QPSK	3/4	78.00	86.80	162.00	180.00
27	4	16-QAM	1/2	104.00	115.60	216.00	240.00
28	4	16-QAM	3/4	156.00	173.20	324.00	360.00
29	4	64-QAM	2/3	208.00	231.20	432.00	480.00
30	4	64-QAM	3/4	234.00	260.00	486.00	540.00
31	4	64-QAM	5/6	260.00	288.80	540.00	600.00

Table 12.6 Comparison of IEEE 802.11n and IEEE 802.11ac

	802.11n	802.11ac
Channel bandwidth	20 and 40 MHz	80 and 160 MHZ
Frequency band	2.4 and 5GHz	5 GHZ
Modulation methods	BPSK, QPSK 16QAM, and 64QAM	Same as 802.11n plus 256QAM
MIMO	4*4	8*8
Transmission	Single client	Multiple Clients
Support MAC combination	A-MSDU, AMPDU	A-MSDU, AMPDU
Maximum, data rate	>600 Mbps	> Gigabits (80 MHz Channel)

12.15   IEEE 802.11AC

IEEE 802.11ac is an improvement of 802.11n in the 5 GHz band. The IEEE802.11ac specification was certified by the Wi-Fi Alliance in mid-2013, and it is referred to as very high throughput (VHT).

The IEEE 802.11AC also adds new technology to the MIMO called Multiuser MIMO. The MIMO increases the data streams for a single client, but Multiuser MIMO enables the Access Point to stream data to multiple clients. The following table compares 802.11n with 802.11ac (Table 12.6).

802.11 g, n, and ac use OFDM and divide the channel into subchannels with a bandwidth of 312.5 kHz for each channel.

Summary

- The standard for wireless LANs is the IEEE 802.11 series.
- Components of WLANs are the WLAN NIC, Access Point, and NOS.
- WLANs uses RF and IR signals for transmitting information.
- The IR ray mode of transmission is used indoors and cannot penetrate through obstacles.
- IEEE 802.11 defines Frequency hopping Spread Spectrum (FHSS) and Direct Sequence Spread Spectrum (DSSS) for the physical layer using RF signals.
- IEEE 802.11 defines Carrier Sense Multiple Access with Collision Avoidance for the access method.
- WLANs use the ISM-band and the U-NII band.
- WLAN topologies are: managed wireless network and unmanaged wireless network.
- Managed Wireless network topologies are: Basic Service Set (BSS) and Extended Service Set (ESS).
- Wireless unmanaged networks are called Ad-Hoc networks.
- Narrow band signal refers to a signal with narrow spectrum.
- Spread spectrum signal refers to a signal with a range of frequencies.
- IEEE 802.11, IEEE 802.11b, and IEEE 802.11g operate in the ISM band, and the data rates for IEEE 802.11b are 1, 2, 5.5, and 11 Mbps.
- IEEE 802.11b offers three nonoverlapping channels; channels 1, 6, and 11.
- IEEE 802.11b uses complementary code keying for the transmission of information.
- The IEEE 802.11 family uses CSMA/CA and point of coordination function (PCF) for the access method.
- IEEE 802.11g operates in 2.4 GHz and uses DSSS and OFDM for the transmission of information.
- The Wireless alliance offers certification for wireless equipment interoperability as referred to Wi-Fi.
- WLAN uses SSID, MAC filtering, WEP, and authentication for security.

Key Terms

Access Point (AP)	Infrared (IR)
Ad-Hoc	Infrastructure band (U-NII)
Authentication	Multipath fading
Basic Service Set (BSS)	Narrowband signal
Beacon frame	Orthogonal Frequency Division Multiplexing (OFDM)
Carrier Sense Multiple Access with	Peer-to-Peer-Network
Cell	Radio frequency (RF)
Chip	Roaming
Collision Avoidance (CSMA/CA)	Service Set Identifier
Complementary Code Keying (CCK)	Spread spectrum
Direct sequence Spread Spectrum (DSSS)	Unlicensed National Information
Extended Service Set (ESS)	Wi-FI
Frequency Hopping Spread Spectrum (FSH)	Wireless local area network
Industrial Scientific and medical Band (ISMB)	

Review Questions

Multiple Choice Questions

1. What is the transmission media for WLAN?

   (a) Air
   (b) Cable
   (c) Optical cable
   (d) UTP

2. An advantage of using spread spectrum signals over narrowband signals is that:

   (a) Spread spectrum has more power.
   (b) Spread spectrum signals use a range of frequencies.
   (c) Narrowband signals use a range of a frequencies.
   (d) Spread spectrum uses a single frequency.

3. The area covered by an Access Point is called a ____.

   (a) frame
   (b) token
   (c) cell
   (d) chip

4. Which of the following technologies are used for WLAN?

   (a) Infrared
   (b) Radio frequency
   (c) a & b
   (d) Digital signal

5. What is the IEEE standard for WLAN?

   (a) IEEE 802.10
   (b) IEEE 802.11
   (c) IEEE 802.12
   (d) IEEE 802.13

6. IEEE 802.11g offers data rates of _____.

   (a) 1 and 2 mbps
   (b) 1, 2, and 11 mbps
   (c) 54 mbps
   (d) 11 and 45 mbps

7. The nonoverlapping channels for IEEE 802.11b are_____.

   (a) 1, 10, and 11
   (b) 1, 6, and 11
   (c) 1, 4, and 10
   (d) 3, 4, and 5

8. IEEE 802.11g operates in the _____ band.

   (a) U-NII
   (b) ISM
   (c) B
   (d) C

9. IEEE 802.11n operates in the _____ bands

   (a) U-NII
   (b) 2.4 GHz
   (c) 5 GHz
   (d) b&c

10. IEEE 802.11g uses _____ and _____for transmitting information.

   (a) DSSS and FHS
   (b) DSSS and OFDM
   (c) DSSS and CCK
   (d) FHS and OFDM

11. A Beacon frame is transmitted by a _____

   (a) client station
   (b) Access Point
   (c) client and Access Point
   (d) distributed system

12. DSSS uses _____chip bits.

   (a) 11
   (b) 12
   (c) 15
   (d) 20

13. IEEE 802.11G uses_____ technology for transmission of information.

   (a) CCK
   (b) OFDM
   (c) FHS
   (d) HR-DSSS

Short Answer Questions

1. What does WLAN stand for?
2. What are the components of a WLAN?
3. What are the WLAN topologies?
4. What is a cell?
5. What is an AD-Hoc topology?
6. Explain narrowband signals and spread spectrum signals.
7. What is the function of an Access Point device in a WLAN?
8. What are the advantages of a spread spectrum signal over a narrowband signal?
9. What are the IEEE standards for WLANs and their data rates?
10. What does OFDM stand for?
11. Explain OFDM operation.
12. What does DSSS stand for?
13. Explain DSSS.
14. What are the data rates for IEEE 802.11b?
15. How many nonoverlapping channels are offered by IEEE 802.11b or g?
16. What is the maximum data rate for IEEE 802.11a?
17. What is the range of frequencies for the U-NII band?
18. What are the types of access methods for WLANs?
19. What are the types of frame formats for WLANs?
20. What is the function of the association request frame?
21. Which devices transmits beacon frames?
22. What is the Service Set Identifier?
23. What does CSMA/CA stand for?

24. What is the function of Wi-Fi?
25. What are the causes of signal distortion?
26. Explain multipath fading.
27. Explain the operation of FHSS.
28. Explain access methods for WLAN.
29. What does ISM stand for?
30. What does MIMO stand for?
31. What is the maximum number of transmitters and receivers for IEEE 802.11n?
32. What are the frequency bands for IEEE 802.11n?
33. What are the channel frequencies in which IEEE 802.11n can operate?
34. List the types of frame aggregation used by IEEE 802.11n.
35. What is the advantage of frame aggregation?
36. What type of frame aggregation is used for single destination and single application?
37. What type of frame aggregation is used for single destination and multiple applications?
38. Does MIMO transmit spatial streams in one frequency or several different frequencies?
39. What is the guard interval?
40. What is application beamforming?
41. What is the maximum data rate of IEEE 802.11n?

Chapter 13
Low Power Wireless Technologies for Internet of Things (IoT)

Objectives

After completing this chapter, you should be able to:

- List types of Low Power Wireless Networks.
- Describe Low Power Wide Area Networks.
- List the applications of ZigBee.
- Explain ZigBee topologies.
- Show the ZigBee Protocol Architecture.
- List ZigBee device types.
- Show the ZigBee physical layer frame format.
- Describe ZigBee node address assignment.
- List ZigBee physical management services.
- Explain the IEEE standard for ZigBee's MAC and physical layers.
- Show the 6LoPAN Protocol Architecture.
- Describe the Application of 6LoPAN.
- List the applications of the LoRa Wide Area Network.
- Describe LoRaWAN characteristics.
- List LoRa WAN components and their function.
- Show LoRaWAN Protocol Architecture.
- Describe LoRaWAN end devices.
- List LoRaWAN security keys.

Introduction

Internet of Things technology (IoT) is growing fast due the advancement of low power wireless sensor networks. These low power networks find use in automation of commercial building systems, home automation, industrial automation, energy and utility automation, healthcare, and remote control. The power of a wireless network can be classified based on the network's distance coverage.

A. Elahi, A. Cushman, *Computer Networks*,
https://doi.org/10.1007/978-3-031-42018-4_13

A. Low Power Wireless networks cover up to 100 m and include Bluetooth, ZigBee, Z-wave, Thread, 6LoWPAN, and WirelessHart.
B. Low Power Wide Area network technologies cover up to 15 km and include LoRa, Sigfox, and DASH.

The remainder of this chapter will cover ZigBee, 6LoWPAN, and LoRa in detail.

13.1   ZigBee

In 2007, the ZigBee Alliance published two feature sets called ZigBee and ZigBee PRO. The ZigBee has the following characteristics:

- Low battery consumption: a ZigBee end device should operate for months or even years without battery replacement.
- Low data rate: the maximum data rate for ZigBee device is 250 Kbps.
- Easy to implement.
- Supports up to 65,000 nodes connected in a network.
- ZigBee can automatically set up its network.
- ZigBee uses small packets, unlike Wi-Fi and Bluetooth

Table 13.1 shows a comparison of ZigBee with Bluetooth and Wi-Fi.

13.1.1   ZigBee Operation and Components

ZigBee Node Each ZigBee node consists of a microcontroller containing the ZigBee protocol stack and a transceiver, as shown in Fig. 13.1.

Table 13.1 Comparison of ZigBee with Bluetooth and Wi-Fi

	Wi-Fi IEEE 802.11n	Bluetooth IEEE 802.15.1	ZigBee IEEE 802.15.4
Application	Wireless LAN	Cable replacement	Control & monitor
Frequency bands	2.4 GHz	2.4 GHz	2.4 GHz, 868 MHz, 915 MHz
Battery life (days)	0.1–5	1–7	100–7000
Nodes per network	30	7	65,000
Bandwidth	2–600 Mbps	3 Mbps	20–250 Kbps
Range (meters)	1–100	1–10	1–75 and more
Topology	Tree	Tree	Star, tree, cluster tree, and mesh
Standby current	$20*10^{-3}$amps	$200*10^{-6}$amps	$3*10^{-6}$amps
Memory	100 KB	100 KB	32–60 KB

Fig. 13.1 ZigBee node

ZigBee Node Operation   The ZigBee node can operate as a full function device (FFD) or reduced function device (RFD) for specific operations. The full function device operates in the full IEEE 802.15.4 MAC layer, while the reduced function device performs only a limited number of tasks.

ZigBee Device Types   ZigBee offers following device types:

A. **Coordinator**: A coordinator is a FFD and is responsible for overall network management. Each network has exactly one coordinator. The coordinator performs the following functions:

   • Selects the channel to be used by the network.
   • Starts the network.
   • Defines how addresses are allocated to nodes or routers.
   • Permits other devices to join or leave the network.
   • Holds a list of neighbors and routers.
   • Transfers application packets.

B. **Router**: A router is a FFD which is used in tree and mesh topologies to expand network coverage. A router performs all functions like a coordinator except for the establishing of a network.

C. **End Device**: An end device is a RFD that performs the following functions:

   • Each end device (child) can be connected to a router or coordinator (parent).
   • Can join or leave a network at will.
   • Transfers application packets.

D. **ZigBee Trust Center (ZTC)**: The ZigBee Trust Center is a device which provides security management, security key distribution, and device authentication.

E. **ZigBee Gateway**: The ZigBee Gateway is used to connect the ZigBee network to another network, such as a LAN, by performing protocol conversion.

13.1.2   ZigBee Topologies

ZigBee offers Star, Tree, and Mesh topologies. Figure 13.2a–c all show ZigBee topologies. Note that each topology uses only one controller.

13.1.3   ZigBee Application Profiles

The ZigBee alliance developed several application profiles that are ready to be used by the users, which are as follows:

- Home automation (HA)
- Smart Energy (SE)
- Commercial Building Automation (CBA)
- ZigBee Health Care (ZHC)
- Telecom Applications (TA)
- ZigBee RF4CE Remote Control

One the most popular ZigBee profiles is Smart Energy (SE). The Smart Energy network is a combination of Advanced Metering Infrastructure (AMI) and Home

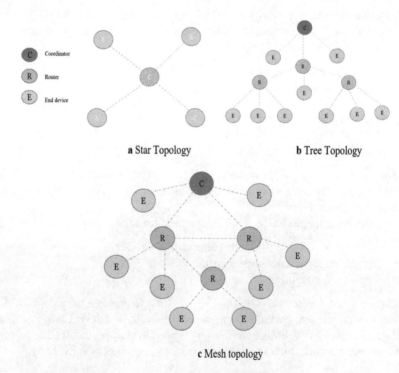

Fig. 13.2 (a) Star topology. (b) Tree topology. (c) Mesh topology

Area Network (HAN), where each meter is equipped with ZigBee node. The AMI network is connected through a gateway and to a server. The Home Area Network consists of a connected washer, dryer, thermostat, and display. The Smart Energy network can perform the following functions:

- The utility company can read the meters at any time.
- Smart energy enables the utility companies to give specific information to customers regarding how to save energy.
- Detects power interruption and location.
- The meter can receive commands and, in the case of emergency, turn off customer power.
- Notify customer of high peak, allowing the customer to reduce consumption.
- Customers can receive time-based pricing, allowing them to make smart energy choices.
- Time-based billing report (monthly, weekly or daily).
- Communicates in two ways: from meter to central utility and from central utility to meter.

Another application profile is the RF4CE Remote control. Radio Frequency for Consumer Electronics (RF4CE) is a protocol developed by a consortium of companies as such as Freescale, Texas Instruments, OKI, Panasonic, Philips, Samsung, and Sony. It defines a standard specification for designing remote control devices for the TV, VCR, and DVD player. The characteristics of a RF4CE device are as follows:

- Does not require line-of-sight to the receiver.
- Supports two-way RF communication between the controller node and the target device.
- Communication between target devices.
- Enables the remote control to display device status.
- Supports paging to locate the remote control.
- Operates in 2.4 GHz.
- Uses a multi-star topology.

13.1.4 ZigBee Protocol Architecture

Figure 13.3 shows the ZigBee Protocol Architecture. The ZigBee Alliance developed the ZigBee Device Object (ZDO), the Application Support Sub-layer (APS), the Network Layer, and Security Management. IEEE 802.15.4 is used for the MAC layer and physical layer.

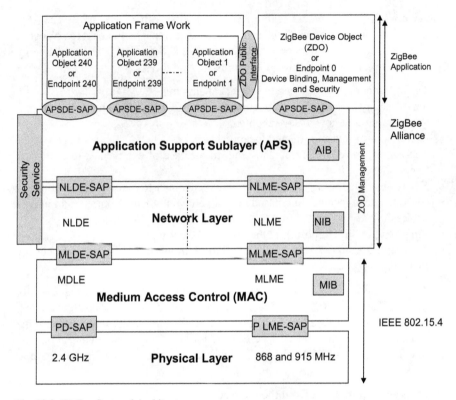

Fig. 13.3 ZigBee Protocol Architecture

13.1.5   Physical Layer

The physical layer performs data service and physical layer management. The functions of the data service are as follows:

- Receiving frames from the upper layer and converting to RF signals for transmission.
- Receiving RF signals from the air and converting them to bits for transfer to the upper layer.

The physical layer management services are as follows:

- Activation and deactivation of the transceiver.
- Clear Channel Assessment (CCA): Checks if the channel is clear or not.
- Energy Detection (ED): Measures energy level of the channel.
- Link Quality Indication (LQI): Indicates the quality of incoming packets.
- Channel Selection: As channels are divided into pages, IEEE 802.15.4 offers 27 channels on page 0 and 11 channels each on pages 1 and 2.

Table 13.2 IEEE 802.15.4 Channels

Page number	Frequency band	Data rate	Modulation	Channel number
0	868 MHz	20 Kbps	BPSK	0
0	915 MHz	40 Kbps	BPSK	1 through 10
0	2.4 GHz	250 Kbps	O-QPSK	11 through 26
1	868 MHz	250 Kbps	ASK	1
1	915 MHz	250 Kbps	ASK	1 through 10
2	868 MHz	100 Kbps	O-QPSK	0
2	915 MHZ	250 Kbps	O-QPSK	1 through 10
3–31	Reserved			Reserved

The IEEE 802.15.4 uses 32 bits to represent the page number and channel number, where the five most significant bits represent the page number and the 27 least significant bits represent the channel number. Table 13.2 shows examples of page numbers and the channel numbers related to each page, as well as the frequency bands, data rate, and the types of frequency modulations. As shown in the table, there are 27 channels in page 0, and on page 2 the type of modulation used is O-QPSK (Offset-QPSK).

The physical layer uses Direct Sequence Spread Spectrum (DSSS) for transmission of information.

Physical Layer Frame Format Figure 13.4 shows the physical layer frame format. The physical layer can carry only 127 bytes from the MAC layer.

13.1.6   IEEE 802.15.4 Medium Access Control (MAC Layer)

The MAC layer performs data services (transmits and receives frames from the upper and lower layers) and management services. The management services perform following functions:

- End device association and disassociation.
- In a coordinator, it offers optional Guarantee Time Slot (GTS) for each device accessing the network.
- In a coordinator, it generates the beacon frame.
- Provides Carrier Sense Multiple Access with Collision Avoidance (CSMA/CA) as the access method for the network.
- Provides reliable connection between two MAC layers by using an acknowledgment frame.

Device Address and PAN ID Each node is assigned a unique address which can either be a 64-bit address assigned by the IEEE or a 16-bit short address. Each network is assigned a 16-bit PAN ID (Personal Wireless Network ID).

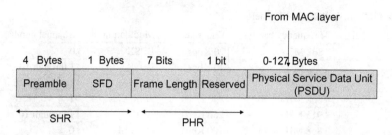

Fig. 13.4 Physical layer frame format

MAC Layer Scanning Channels The MAC layer can perform the following channel scanning operations:

- **Energy Detection**: Energy detection is used by a coordinator to measure the energy level of selected channels in order to select the best channel.
- **Active Scan**: The active scan is used to determine if any network is in its vicinity. This is done by sending a beacon request command.
- **Passive Scan**: The full functional device or reduced functional device listens for any beacon transmitted by any coordinator. A passive scan is used by a device in order to join a network.
- **Orphan Scan**: Orphan scan is used by a device which has lost its parent and is trying to re-associate itself with the parent device.

General MAC frame format Figure 13.5 shows the MAC frame format and Fig. 13.6 shows the frame control fields.

The FCS is calculated over the MHR and MAC payload parts. FCS uses CRC $G(16) = \mathbf{X^{16} + X^{12} + X^5 + 1}$

Frame Type 000 denotes a Beacon Frame, 001 a Data Frame, 010 an Acknowledge Frame, 011 a Command Frame, and 100–111 are all reserved.

PAN ID Compression This bit indicates if the destination device is located in the same network or a different network as the source device.

Source and Destination Addresses Mode 00 means indirect addressing, 10 means 16-bit addressing, and 11 means 64-bit addressing.

Frame Pending This field indicates there is more data being held for this device which the device can request.

Auxiliary Security ZigBee does not support MAC layer security.

13.1.7 ZigBee Network Layer

The Network Layer performs Data and Management services. Management services are as follows:

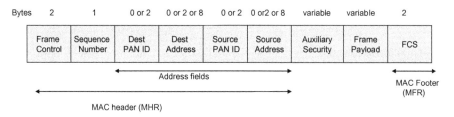

Fig. 13.5 MAC frame format

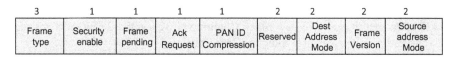

Fig. 13.6 Control field

- **Configuration of new devices**: Configures a new device in order to be the coordinator or simply as a device joining to the network.
- **Starting a network**: Only performed by the network's coordinator.
- **Addressing**: The coordinator and routers can assign addresses to each end device joining the network.
- **Neighbor Discovery**: Discovers one hop neighbors, recording their addresses and capabilities.
- **Route Discovery**: Finds the most efficient route over which to transfer messages to a destination.
- **Discover other networks.**
- **Security**: Applies the security protocol to outgoing and incoming frames.
- **Joining and leaving a network**
- **Establishing Network Topologies**: Star, Mesh or Tree.
- **Address Assignment**: Stochastic Address Assignment is used by ZigBee Pro. ZigBee Pro uses a Mesh Topology and each node when joining the network is assigned a random number from 0 to 65,536. The device will broadcast a device announcement to inform the other devices of its presence in the network. If any device in the network discovers an address conflict with the new device, a conflict notification will be broadcasted, and the parent of the child will assign another address to the child.
- **Distributed scheme (Cskip)**: used by a network with tree topology.

Routing The function of a router in a ZigBee network is to provide a routing protocol so that a message may be routed from a source to a destination. ZigBee supports the following routing protocols:

- Tree Routing
- Ad hoc On-demand Distance Vector (AODV)
- Many-to-one
- Source routing

13.1.8   Application Layer

The application layer consists of application objects (endpoints) which hold user applications and ZigBee Device Objects (ZDOs). Each node has 240 endpoints meaning each node can have 240 sensors.

ZigBee Security ZigBee provides message integrity, authentication, freshness and privacy for a ZigBee device.

- ZigBee security uses Counter Mode (CTR) with 128-bit AES for the encryption of messages.
- ZigBee security uses Cipher Block Chaining (CBC) with 128-bit AES for the generation of the Message Integrity Code (MIC).
- ZigBee security uses symmetric keys for all levels of security.
- ZigBee security can apply cryptography and frame integrity at the application and network layers.

ZigBee Security Keys and Trust Center ZigBee defines three types of keys for security: the link key which is used for the application link key and Trust Center Link, as well as the Network Key and Master Key.

Application Link Key This key is used for security of application data between two devices and it is shared only between two devices. An application link key may be preconfigured, distributed by a trust center to the devices, generated from a master key, or installed using the Symmetric-Key Key Establishment (SKKE) protocol.

Trust Center Link Key This key is used by the trust center and devices on the network for securing communication between the trust center and devices. This key is preconfigured in the devices.

13.1.9   ZigBee Security Modes

Standard Mode Standard security mode is used by ZigBee and ZigBee Pro. Standard security uses two network keys that are transmitted by the trust center to the devices for encryption and decryption.

High Security Mode High security mode is used by ZigBee Pro. The high security mode provides all functions supported by standard security with the following additional functions:

- **Entity authentication**: This is used by two devices to authenticate each other based on their active network key.
- **Permission table**: This table indicates which devices have permission for using commands such as permission to join or leave.
- Generating link keys between devices using the SKKE protocol.
- Two network keys
- The trust center holds network keys, master keys, and list of devices

13.2 6LoWPAN Architecture

6LoWPAN stands for IPV6 over Low Power Personal Area Network. 6LoWPAN was developed by the IETF (Internet Engineering Task Force) to enable Low Power Wireless Devices to be able to carry IPv6 packets in order to support the Internet of Things. 6LoWPAN defines encapsulation and header compression mechanisms that allow IPv6 packets to be sent and received over IEEE 802.15.4-based networks. Assume in Fig. 13.5 that the Auxiliary Security field is 8 bytes, then the size of MAC header will be 31 bytes. Add to that the size of the MAC Header and MAC footer = 31 + 2 = 33 bytes.

The Physical Layer Payload of IEEE802.15.4 can carry only 127 bytes; therefore, the maximum size of IPV6 will be 127 − 33 = 96 bytes. From these 96 bytes, 40 bytes are used for IPV6, resulting in 56 bytes used for the TCP packet. The IETF developed 6LoWPAN layered architecture as shown in Fig. 13.7.

The functions of the Adaptation Layer are as follows:

A. **Compression**: The IPv6 header can be compressed from 40 bytes to a minimum of 3 bytes, but the hop limit cannot be compressed. When the Adaptation Layer is set to 01000001, it means that there is no IPV6 header compression, as shown in the uncompressed IPv6 header in Fig. 13.8.

If the Adaptation Layer is set to 01000010, then that means the IPV6 header is compressed based on information provided in the HC1 field, as shown in Fig. 13.9. The HC1 field determines which field of IPV6 is to be compressed.

B. **Fragmentation**: Fragments the IPv6 payload into multiple frames. Figure 13.10 shows a combination of fragmentation with compression.

Figure 13.11 shows the 6LoWPAN architecture where the Low Power Wireless Networks are connected to the Internet through the Border Routers. The Border Routers perform the following functions:

- Compress incoming packets from an IPv6 Network and forward them to 6LoWPAN.
- Segmentation of large IPv6 packets and forward the segments to 6LoWPAN.
- Decompress incoming packets from 6LoWPAN and forward them to an IPv6 Network.

Fig. 13.7 6LoWPAN architecture

6LoWPAN Application
UDP/TCP
IPV6/ ICMP
Adaptation
IEEE802.15.4 MAC
IEEE802.15.4 PHY

40 bytes

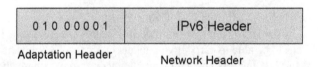

Fig. 13.8 Adaption layer with no compression

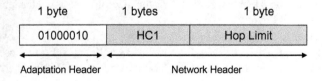

Fig. 13.9 Header compression

802.15.5 MAC Header	Fragment Header	IPv6Compression Header	IPv6 packet

Fig. 13.10 Fragmentation with compression of IPV6

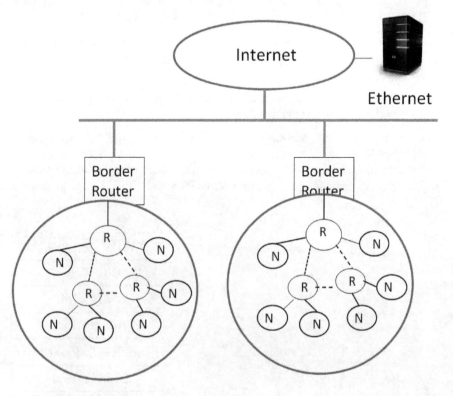

Fig. 13.11 6LoPAN Network

13.3   LoRa Wide Area Network Technology (LoRa WAN)

Low Power Wide Area Network (LPWAN) technology is capturing a large market of IoT, which offers long range, low power, low data rate communications. There are several LPWAN technologies such as LoRa, SIGFOX, and NB-IoT. LoRa stands for Long Range, and it is a wireless modulation method that is used for LPWAN. LoRa WAN wireless technology was developed by the LoRa Alliance and LoRa wireless radio frequency technology was developed by SEMTECH Corporation. Table 13.3 shows some of the applications of LoRa.

13.3.1   LoRa WAN Characteristics and Architecture

The following are LoRa WAN characteristics:

1. LoRaWAN Topology: LoRaWAN is a Star of Star topology network. Several end devices communicate with one Gateway in the form of a Star topology, and Gateways and Network servers also form a Star topology.
2. LoRaWAN operates in unlicensed ISM band.
3. LoRaWAN offers long range coverage, up to 5 km in an urban and 15 km in a suburban environment.
4. LoRaWAN uses low power consumption, 10 years of battery life at +14 dBm (31 mW) transmitter power at 868 MHz.
5. LoRaWAN offers bidirectional communications.
6. LoRaWAN can have up to 300 K Nodes.
7. LoRaWAN offers Adaptive Data Rate (ADR).
8. LoRaWAN offers Authentication and Security.
9. LoRaWAN offers data rates from 300 bps up to 5 kbps at the 125 KHz frequency band and 10 Kbps at the 250 KHz band.

13.3.2   LoRaWAN Components and Functions

LoRa End Device (Sensor Node)   The end device communicates with the Gateway through LoRa RF, and the Gateways communicate with a network server through TCP/IP through a public or private network. The end device transmits packets in the

Table 13.3 LoRa applications

Smart metering	Agriculture	Smart city	Health
Electric meters	Irrigation control	Smart parking	Medical wearable
GAS meters	Environmental sensing	Street lighting	Connected bracelets
Water meters	Animal tracking	Vending machine	Condition monitoring

form of broadcasts and receives from multiple gateways within the range. The gateways then transmit the packets to a network server. With this design, the network server will receive multiple copies of a packet and select one of the packets to transmit to the appropriate application server. For example, if the packet belongs to the gas meter, then the network server will transmit the packet to the Gas Meter Server (an application server).

End Device Identifiers   Each end device has the following identifiers:

A. Device Identifier (DevEUI): This is a 64-bit unique identifier which is set by vendors or developers.
B. End Device Address: Each end device has a 32-bit address which is used for communication between the end device and gateway. The device address is unique in the network and assigned by the network server to the device during the join process.
C. Application Identifier (AppEUI): AppEUI is a unique identifier for the application server that assigns to the end device. This is used by the network server to send the packet to the appropriate application server.

LoRa Gateway   The LoRaWAN gateway is connected to the network server through Ethernet or the Internet. The end device transmits the packet to gateway, then the gateway demodulates the received signals, adds the following information to the packet, and then transmits to the network server:

(a) The arrival time of the packet
(b) SNR (Signal to Noise Ratio)
(c) RSSI (Receive Signal Strength)
(d) Data rate
(e) Channel number

The gateway listens to multiple channels and processes multiple packets. In a LoRaWAN with multiple gateways, the end device broadcast message may be received by multiple gateways. The gateways transmit the packets to the network server, then the network server will select the best packet based on signal quality and reject rest of the packets.

Network Server   The network server performs the following functions:

A. The network server receives the data packets from multiple gateways and removes duplicate packet if received by multiple gateways then transmit one packet to the application server.
B. The communication between the Gateways and application servers can use by Public or Private Networks
C. The network server responsible for managing end devices and Gateways such as device Join request, device authentication, assign an address to the end device, and register new device.
D. The Network Server supports channel re-configuration. It will automatically disable channels on the node that are not being used by the network.

E. The network server can implement Adaptive Data Rate (ADR) for the end device. The ADR is used to assign the best data rate to end device with the least transmission power.

F. The network server uses information when receiving the packet from the gateway, such as the SNR or SSR to select the best gateway for the downlink transmission.

Application Server The application server is connected to the network server. It decrypts incoming packets and encrypts outgoing packets using the Application Session Key.

Join Server The function of the join server is to activate and de-activate the end device, as well as to also generate session security keys.

13.3.3   LoRaWAN Protocol Architecture

The LoRa Alliance developed the LoRa Protocol Architecture. Figure 13.12 shows the LoRaWAN Protocol Architecture, which consists of a Physical, Data Link, and Application Layer. The word "LoRa" refers to the Physical (PHY) layer of LoRaWAN.

The physical layer consists of LoRa Modulation with regional ISM bands. The LoRa physical layer uses Chirp Spread Spectrum (CSS) modulation which enables LoRaWAN to consume low power and transmit rf signals up to 15 km.

LoRa MAC Layer
The LoRa MAC Layer offers three different types of end-user operations, those being Class A, Class B, and Class C. The LoRa MAC Layer performs the following functions:

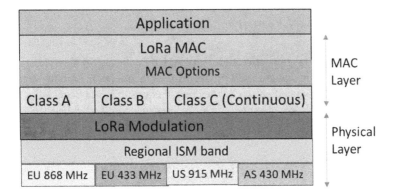

Fig. 13.12 LoRaWAN Protocol

A. Determine end device type and its operation
B. Defines access method
C. Generates security keys
D. Defines Adaptive Data Rate (ADR)

LoRaWAN defines the three types of end device operations as Class A, Class B (Beacon), and Class C (Continuous)

Class A Class A supports bidirectional transmission and initiates transmission. Class A devices can transmit at any time but will receive after transmission. The Class A is used for end device transmission with minimum downlink transmission from the server, such as for temperature sensors and meters. Each end-device uplink transmission is followed by two short downlink receive windows, which is shown in Fig. 13.13.

Class B Class B receives a Beacon frame from the gateway which contains the schedule for allocating a time slot for transmission. The end devices of Class B consume more energy than Class A.

Class C End devices of Class C are in receiving mode can receive the packet at any time. They will only switch to transmitting mode when transmitting. Class C end devices will use more power to operate when compared with Class A and Class B.

LoRaWAN Access Method LoRaWAN uses the Aloha Protocol as an access method. In the Aloha Protocol, the device transmits and waits for an acknowledgment. If the device did not receive acknowledgment, then it assumes a collision occurred and the device re-transmits according to a back-off algorithm.

Adaptive Data Rate (ADR) The ADR is a method that can adapt the data rate of the channel based on the strength of radio signal, the S/N. If the S/N is small, then the data rate must be lowered (using a higher Spreading Factor). Selecting a lower SF results in a higher data rate.

13.3.4   LoRaWAN Security

When a node requests connection to LoRaWAN, then the network server and nodes generate several keys that are used for security, authentication, and message integrity. The LoRaWAN specification defines two versions of LoRa Security, version 1, and version 1.1. LoRAWAN version 1 only uses one root key called the APPKey (Application Key) to generate the Application Session Key and Network Session Key for security. To address the shortcomings of security version 1, the LoRa Alliance published LoRaWAN security version 1.1.

Fig. 13.13 Class A operation

LoRaWAN Version 1.1 Security

LoRa defines two methods that an end device can use to join a LoRaWAN, and they are as follows:

A. Over-The-Air-Activation (OTAA)
B. Activation by Personalization (ABP)

Over-The-Air-Activation (OTAA) and Security Keys Generation

The Join process consists of a Join Request sent by an end device to the Network Server, and the Network Server then sends a response to the end device before an end device can join the LoRaWAN network. The end device and Network Server share two root keys called the NewkKey (Network Key) and AppKey (Application Key). These two keys are used to generate the Network Session Keys, Application Session Key, and Joint Session Keys. The following identifications are stored in the end device:

A. **Device Identifier (DevEUI)**: It is a 64-bit unique identifier which is set by vendors or developers.
B. **JoinEUI**: It is a global Application ID using IEEE EUI 64 that identifies the Join Server
C. **Application Keys**: AppKey and NewkKey (Network).

Summary

- ZigBee end devices can operate for months and years without battery replacement.
- The maximum ZigBee data rate is 250 Kbps.
- ZigBee supports Mesh, Star, and Tree topologies.
- ZigBee supports 65,000 nodes per network.
- The ZigBee device types are the end nodes, the coordinator, the router, and ZigBee trust center.
- In a ZigBee network, only one coordinator can exist.
- The function of the coordinator is to manage the network and assign addresses to the end nodes and routers.
- ZigBee defines two types of node operations: full function and reduced function.
- End devices operate at reduced function and the coordinator operates at full function.
- ZigBee developed applications for home automation, Smart Energy, healthcare, and more.
- ZigBee uses IEEE 802.15.4 for the physical and MAC layers.
- ZigBee operates at 2.4 GHz band in USA.
- ZigBee end devices use CSMA/CD to access the network.
- ZigBee offers Standard and High security modes.
- ZigBee operates at a short-range distance of 100 m.
- 6LoWPAN stands for IPV6 over Personal Area Network.

- IETF developed 6LoPAN in order to allow Low Power Wireless Devices to connect to the Internet using IPv6.
- IEEE802.15.4 can carry only 137 bytes.
- LoRa offers long-range communication up to 5 km in urban and 15 km in suburban environments, called a LoRa Wide Area Network.
- LoRa stands for Long Range.
- LoRa applications are electric meters, irrigation control, water meters, and more.
- LoRa uses the star of star topology.
- LoRa operates in the ISM band.
- LoRa uses frequency modulation.
- LoRa offers a data rate between 300 bps and 5 Kbps.
- LoRa components are LoRa end devices, LoRa Gateway, and Network Server.

Key Terms

Active Scan	LoRa WAN
Adaptive Data Rate (ADR)	LoRaWAN Access Method
Application Link Key	Orphan Scan
Border Router	Passive Scan
Commercial Building Automation (CBA)	Over-The-Air-Activation (OTAA)
Coordinator	Smart Energy (SE)
Distributed scheme (Cskip)	Trust Center Link Key
End device	Telecom Applications (TA)
Energy Detection	ZigBee
Home automation (HA)	ZigBee Health Care (ZHC)
IEEE802.14.5	ZigBee Protocol Architecture
IETF	ZigBee RF4CE Remote Control
Low Power Wireless Technologies	6LoWPAN

Review Questions

Multiple Choice Questions

Short Answer Questions

1. List the name of the Low Power Wireless Network that covers up to 100 m.
2. List the name of the Low Power Wide Area Network Technology that covers more than 100 m.
3. What is the maximum data rate for ZigBee?
4. What is the maximum number of nodes that can be used in a ZigBee network?
5. What is the frequency band for ZigBee in the USA?
6. List the ZigBee topologies.
7. What are the components of a ZigBee device?

8. List ZigBee device types and explain the function of each device.
9. List the applications of ZigBee.
10. What is IEEE 802.15.4?
11. Show the ZigBee Protocol Architecture.
12. List the functions of the ZigBee physical layer.
13. How many channels does ZigBee Support on page 0?
14. Show the ZigBee physical layer frame format.
15. What is the maximum payload for ZigBee?
16. How many bits are addressed to a ZigBee end device?
17. List three functions of the ZigBee MAC layer.
18. What is the difference between a full function device and a reduced function device?
19. What is the function of the Coordinator in a ZigBee network?
20. What method does ZigBee use for encryption?
21. What method does ZigBee use for message integrity?
22. List the ZigBee security modes.
23. What does 6LoWPAN stand for?
24. What is the application of 6LoWPAN?
25. Show the 6LoWPAN Protocol Architecture.
26. What is the function of the Adaptation Layer in 6LoWPAN?
27. What is the function of the Border Router in 6LoWPAN?
28. List four applications for LoRa WAN.
29. What is the topology of LoRa WAN?
30. What frequency band does LoRa operate?
31. What is the maximum coverage of LoRa WAN?
32. What is the data rate of LoRa WAN?
33. List LoRa WAN components
34. Show the LoRa WAN Protocol Architecture

Chapter 14
Introduction to Cryptography

Objectives
After completing this chapter, you should be able to:

- Explain different types of network attacks.
- Define the elements of network security.
- Understand the basics of cryptography.
- List encryption algorithms.
- Distinguish between symmetric and public key cryptography.
- Explain how Diffie–Hellman generates symmetric key.
- Explain how elliptic curve is used for cryptography.
- Understand the application of digital certificates and signatures.

Introduction
Cryptography is a technique used to establish secure communications and is used frequently in the case of network security. Information flowing across public spaces such as the internet drives a need for encrypted communications to prevent eavesdropping or any other malicious activity. Additionally, cryptography can assist in verifying the integrity of data to ensure that it is not modified or otherwise altered in transit. There are two main types of network attacks:

(i) **Direct Attack:** The attacker is able to disrupt the system by breaking passwords accessing the system to modify information. For example, a person breaks the security of a bank server and then alters the account information.

(ii) **Indirect Attack or Passive Attack:** The attacker obtains the information and data in a system (such as a name, address, social security, or credit card number). For example, a person using a cable modem can use a packet sniffer to capture packets transmitted over the modem and obtain sensitive information. A packet sniffer is a piece of software that captures packets traveling in and out of a network. The packet sniffer is used for network monitoring or analysis. The following are some common attacks used against an organization's network:

A. Elahi, A. Cushman, *Computer Networks*,
https://doi.org/10.1007/978-3-031-42018-4_14

- **Eavesdropping or Packet Sniffing**: The attacker intercepts network data or authentication information.
- **Impersonation**: The attacker will impersonate a user and create unauthorized content such as email messages.
- **Denial of Service (DoS):** Make the network resources unavailable.
- **Replay Messages**: Gain access to a data packet, change information within it a then retransmit the falsified information.
- **Password Cracking**: Breaking the password by using dictionary or brute force attacks.
- **Guessing of Encryption Key**: Guessing encryption key in order to decrypt information.
- **Port Scanning:** To find open and insecure ports and services which may be exploited.

14.1 Elements of Network Security

In order to have secure network, the following network security components are necessary:

 (i) **Confidentiality**: Confidentiality, or secrecy, means that information being sent through the network must remain unknown to unauthorized people. This is handled through the use of various encryption methods.
 (ii) **Authentication:** Authentication methods verify the identity of a person or computer accessing the network.
(iii) **Authorization:** To prevent unauthorized access.
(iv) **Integrity:** Integrity maintains data consistency and prevents tampering with information.
 (v) **Non-repudiation**: Non-repudiation provides proof of origin to the recipient.

14.2 Introduction to Cryptography

Cryptography is the analysis and deciphering of codes and ciphers. In computer networking, cryptography is the science of keeping information secure. The process of encoding and decoding information is called encryption and decryption respectively. Figure 14.1 shows a cryptographic model.

In this model, the message is referred to as **plaintext** or **cleartext** and the encrypted text is called **ciphertext**. The plaintext is encrypted by an encryption algorithm and an encryption key. The ciphertext is then transmitted over the communication channel. At the receiving side, the ciphertext is decrypted by a decryption algorithm and a decryption key resulting in plaintext. The encryption and decryption algorithms are called **Ciphers**. Cryptanalysis is the art of breaking

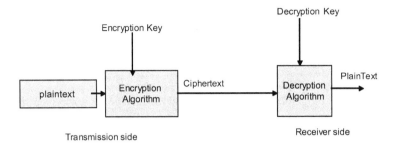

Fig. 14.1 Cryptographic model

ciphers or decrypting information without having the key. Cryptography can be divided into two classes:

A. **Classical Cryptography**: Classical cryptography was used in the early days for non-computing applications such as character substitution and transposition ciphers.

- *Substitution Cipher*: Substitution replaces each character in the text with a different character. This is an early cryptography method that was developed by Caesar and is known as the *Caesar Cipher*. Caesar Cipher substitutes the original letter with a different letter that comes later in the order of the alphabet. For example, the letter "A" is replaced with the 4th letter in the alphabet, substituting "A" with "E." Another method of substitution is to add a number to the ASCII value of the letter. The ASCII value of "A" is 41h. Adding 4 to 41h results in 45h (the letter "E"). The number 4 is the value of the key.

Example 14.1 Encrypt the word "BOOK" assuming the encryption key is 4.

The word "BOOK" in ASCII code is 41, 4F, 4F, 4B. Adding 4 to each character results in 45, 53, 53, 4F which stands for "ESSK."

- *Transposition Ciphers*: In a transposition cipher, the plaintext is divided into groups of characters where the number of groups defines the key. For example the phrase "WELCOME TO CLASS" is divided into groups of 4 characters as shown below.

1	2	3	4
W	E	L	C
O	M	E	–
T	O	–	C
L	A	S	S

- Reading from column 1 to 4, the ciphertext is "WOTLEMOALE-SC-CS." In this example the key is 1234. Also note that the order of the key can be changed to 2341.

B. **Modern Cryptography**. Modern cryptography is now used for modern data communication. The types of modern cryptography are **symmetric key cryptography** (private key or secret key) **and asymmetric key cryptography** or **public key cryptography**.

Symmetric Cryptography

In symmetric cryptography, the transmitter and the receiver of a message share the same key for encryption and decryption as shown in Fig. 14.2.

Plaintext is encrypted using an encryption algorithm and a key, resulting in ciphertext which is then transmitted to the receiver. The receiver uses the same key as the transmitter to decrypt the ciphertext into plaintext. Some of the symmetric cryptographies are **RC4, Data Encryption Standard (DES), Blowfish, and Advanced Encryption Standard (AES)**

The advantages of symmetric cryptography are that it is simple and fast. The disadvantage is that the transmitter and the receiver must possess the same key, meaning that there must be some sort of transfer mechanism. The symmetric algorithm can be divided into **stream ciphers** and **block ciphers**.

Stream Ciphers The plaintext is encrypted one bit at a time, text is converted to binary, and then the binary string is XORed with another binary string (the key), resulting in ciphertext. An example of a stream cipher is the **RC4 algorithm.**

Example 14.2

The word "BOOK" in binary form is	1000010 1001111 1001111 1001011
The encryption key is	0111111 0111110 0111010 1010101

Figure 14.3 shows the stream cipher and output of XOR is ciphertext. The ciphertext is transmitted to the receiver which uses the same key to decrypt the text.

Block Ciphers A block cipher is a group of bits that are encrypted. Some of the symmetric cryptography algorithms using block cipher are the Data Encryption Standard (DES) that takes 64 bits per block and the Advanced Encryption Standard (AES).

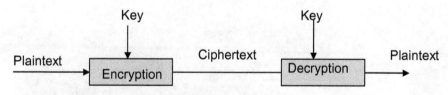

Fig. 14.2 Block diagram of symmetric cryptography

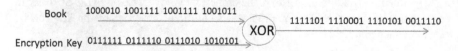

Fig. 14.3 Stream cipher operation

14.3   RC4 Algorithm

RC4 was developed by Ron Rivest and it is also called Ron's Code. RC4 is a symmetric key stream cipher which encrypts bit by bit. The RC4 encryption algorithm consists of two sub-algorithms, and they are as follows:

A. Key Scheduling Algorithm (KSA)
B. Pseudo Random Generation Algorithm (PRGA)

Key Scheduling Algorithm (KSA)
KSA consists of two sub-algorithms, initialization of State Array or S-box and scrambling of State Array.

A. **Initialization State Array**

The RC4 algorithm uses an array of 256 elements which is represented by S[I] where the value of I goes from 0 to 255 and is initialized to a state where S[I] = I.

Initialization Algorithm

```
For I= 0 to 255
S[I]  =  I means  S[0]  =  0,  S[1]  =  1,  S[2]  =  0,  S[3]  =  0,  ….
S[255]  =  255,
or it can represent S[I] in the following form:
S =[0,  1,  2,  3,  …….255]
```

B. **Scrambling of Key Scheduling Algorithm (KSA)**

The encryption key is represented by K[I] and it is L bytes long. The scrambling KSA is represented by:

```
Set J=I
 For I = 0 to 255     then
J= (J+ S[I] + K[I modulo L]) ( modulo N)
 where N represents the number of elements in S[I] and L is the key
 length in bytes.
 Swap S[I], S[J]
```

Example   Assume S = [0, 1, 2, 3] which holds 4 values and each value represented by three bits rather than 8 bits for simplicity

Assume encryption K = [1, 5, 2, 4] and apply the scrambling algorithm:

Fig. 14.4 RC4 generation

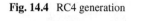

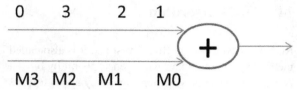

M3 M2 M1   M0

```
Iteration 0 for I=0, J=0
  J = ( 0 + 0  + K( 0  modulo 4) ) modulo 4) = ( 0 + K[0] ) modulo
4  = ( 0 + 1 ) modulo 4 = 1
  Swap S[0]  with S[1]  result   S= [ 1 , 0, 2 , 3]
Iteration 1    for   I=1, J=1, and   S = [1, 0, 2, 3]
  J= (J+ S[I] + K[I modulo L]) ( modulo 256)
  J= (1 + 0 + K[ 1 modulo 4]  (modulo 4) =  (1 + k[1]) modulo 4 =
(1 +5 ) modulo 4 = 2
Swap S[1] with S[2]  result S =[ 1, 2,0, 3]
Iteration 2:  for I=2, J=2, and S = [1, 2, 0, 3]
  J = (2 + 2 + K[ 2 Modulo 4]) modulo 4  = (4 + 2)modulo 4 = 2
  Swap S[2], with  S[2]
Iteration 3:   I=3, J=3   , S=[1,2,0,3], K = [1, 5, 2, 4]
  J= (J+ S[I] + K[I modulo L]) ( modulo N)   =   (3 + 3+ 0)
modulo 4= 2
  Swap S[3] with S [2], then the result is S =[1,2,3,0]
```

If message represented by M = M0, M1, M2, M3, then the cipher text is the output of Fig. 14.4.

14.4   Data Encryption Standard

Figure 14.5 shows a block diagram of DES. DES is a block cipher, where a message is divided into 64-bit blocks of plaintext ($b_0...b_{63}$) and utilizes an initial key of 56 bits. The function of the permutation box is to change the order of bits in the plaintext. For example, bit b_0 becomes b_{58}, b_1 becomes b_{50}, etc. The function of the key generator is to generate 16 different 48-bit keys from a 56-bit key. As shown in Fig. 14.5, the plaintext goes through 16 iterations using different keys and the results are permutated to generate the ciphertext.

Figure 14.6 shows a diagram of a single iteration of DES. The output of the permutation box is divided into two groups of 32 bits, called L_0 and R_0. R_0 is changed to 48 bits by using the expansion/permutation table. The output of the expansion/permutation box is XORed with the 48-bit key creating a 48-bit result. The 48 bits

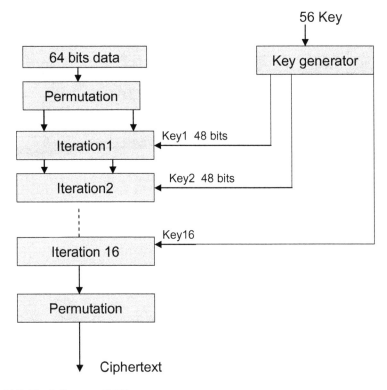

Fig. 14.5 Block diagram of DES

are converted by the substitution box (S-BOX is a table used for converting 48 bits to 32 bits) to 32 bits. These 32 bits are XORed with L_0 yielding the result R_1 for the next iteration. The output of R_0 becomes L_1 for the next iteration.

Key generation is done by dividing 56 bits into two groups of 28 bits. Both groups of 28 bits are shifted by one bit and the output becomes the input to the permutation table where 48 bits are generated for the first key. The following method is repeated with 16 different keys to produce ciphertext as shown in Fig. 14.6.

Triple DES (3DES)
Triple DES is similar to DES but applied three times in series as shown in Fig. 14.7. The 3DES offers following options:

A. All three keys are independent.
B. K1 and K2 are impendent and K3 = K1.
C. All three keys are the same K1 = K2 = K3.

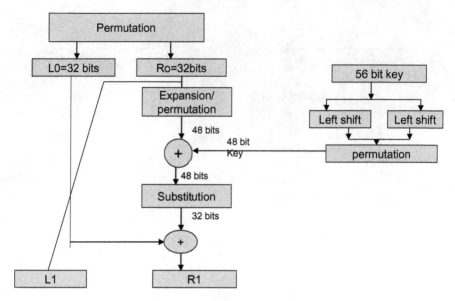

Fig. 14.6 Block diagram of one iteration of DES

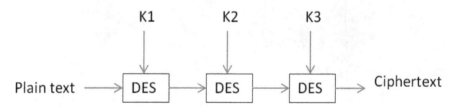

Fig. 14.7 3DES encryption algorithm

14.5   Advanced Encryption Standard (AES)

AES, or Rijndael (the combination of the names Rijman and Daemen, the developers of the AES algorithm), is a block cipher which can use a 128-bit block of plaintext and either a 128-, 192-, or 256-bit key for encryption. Figure 14.8 shows a block diagram of AES. The key generator generates different key for each round.

The number of the rounds is dependent on the key size, and Table 14.1 shows the number of rounds for a given key size.

The cipher text and key are each represented by a 4 * 4 array of bytes (16 bytes * 8 = 128 bits) called the cipher state and key, respectively. Figure 14.9 shows the cipher state and key.

Pre-round Transformation   In pre-round transformation, each element of ciphertext is XORed with the encryption key and the result of the Metrix is input to the first round.

128, 192,or 256 bits
128 bits Plaintext
Per-Round Transformation
K0
Round1
K1
Key Generator
Round 2
K2
Round N
KN
128 bits Ciphertext

Fig. 14.8 Block diagram of AES

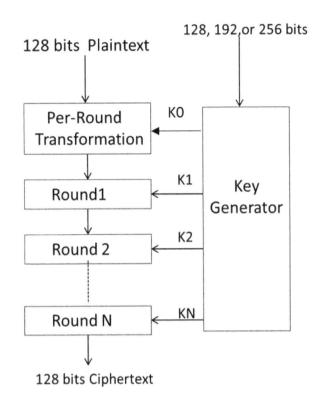

Table 14.1 AES rounds and key sizes

Number of rounds	Key size
10	128 bits
12	192 bits
14	256 bits

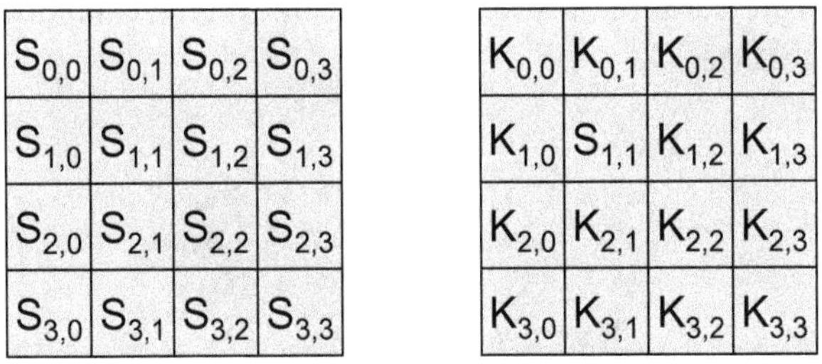

$S_{0,0}$ $S_{0,1}$ $S_{0,2}$ $S_{0,3}$
$S_{1,0}$ $S_{1,1}$ $S_{1,2}$ $S_{1,3}$
$S_{2,0}$ $S_{2,1}$ $S_{2,2}$ $S_{2,3}$
$S_{3,0}$ $S_{3,1}$ $S_{3,2}$ $S_{3,3}$
$K_{0,0}$ $K_{0,1}$ $K_{0,2}$ $K_{0,3}$
$K_{1,0}$ $S_{1,1}$ $K_{1,2}$ $K_{1,3}$
$K_{2,0}$ $K_{2,1}$ $K_{2,2}$ $K_{2,3}$
$K_{3,0}$ $K_{3,1}$ $K_{3,2}$ $K_{3,3}$

Fig. 14.9 Representation of cipher state and key

Table 14.2 S-box table used in AES substitution step

```
   | 0   1   2   3   4   5   6   7   8   9   a   b   c   d   e   f
---|--|--|--|--|--|--|--|--|--|--|--|--|--|--|--|--|
00 |63  7c  77  7b  f2  6b  6f  c5  30  01  67  2b  fe  d7  ab  76
10 |ca  82  c9  7d  fa  59  47  f0  ad  d4  a2  af  9c  a4  72  c0
20 |b7  fd  93  26  36  3f  f7  cc  34  a5  e5  f1  71  d8  31  15
30 |04  c7  23  c3  18  96  05  9a  07  12  80  e2  eb  27  b2  75
40 |09  83  2c  1a  1b  6e  5a  a0  52  3b  d6  b3  29  e3  2f  84
50 |53  d1  00  ed  20  fc  b1  5b  6a  cb  be  39  4a  4c  58  cf
60 |d0  ef  aa  fb  43  4d  33  85  45  f9  02  7f  50  3c  9f  a8
70 |51  a3  40  8f  92  9d  38  f5  bc  b6  da  21  10  ff  f3  d2
80 |cd  0c  13  ec  5f  97  44  17  c4  a7  7e  3d  64  5d  19  73
90 |60  81  4f  dc  22  2a  90  88  46  ee  b8  14  de  5e  0b  db
a0 |e0  32  3a  0a  49  06  24  5c  c2  d3  ac  62  91  95  e4  79
b0 |e7  c8  37  6d  8d  d5  4e  a9  6c  56  f4  ea  65  7a  ae  08
c0 |ba  78  25  2e  1c  a6  b4  c6  e8  dd  74  1f  4b  bd  8b  8a
d0 |70  3e  b5  66  48  03  f6  0e  61  35  57  b9  86  c1  1d  9e
e0 |e1  f8  98  11  69  d9  8e  94  9b  1e  87  e9  ce  55  28  df
f0 |8c  a1  89  0d  bf  e6  42  68  41  99  2d  0f  b0  54  bb  16
```

The following steps describe the operation of each AES round:

1. **Substitution Step**: Each byte in the state is replaced by another byte using an S-box (Table 14.2). For example, if the current byte state is 0x19, it is replaced with the number located at row 0x10 and column 0x09 which is 0xd4.

2. **Shift Row Step**: Performs the following operation on each row of the result of step 1 as shown in Fig. 14.10.

   (a) No shift on first row.
   (b) Circular shift left one byte (each bit is shifted 8 times) on the second row.
   (c) Circular shift two bytes (each bit is shifted 16 times) on the third row.
   (d) Circular shift three bytes (each bit is shifted 24 times) on the fourth row.

3. **Mix Column**: As shown in Fig. 14.11, the result of step 2 is represented by matrix S where the given matrix is multiplied by each column of matrix S to produce matrix A.

   Where

   $$A_{0,0} = 2S_{0,0} + 3S_{1,1} + S_{2,2} + S_{3,3}$$
   $$A_{0,1} = 1S_{0,1} + 2S_{1,2} + 3S_{2,3} + S_{3,0}$$
   $$A_{0,2} = S_{0,2} + S_{1,3} + 2S_{2,0} + 3S_{3,1}$$
   $$A_{0,3} = 3S_{0,3} + S_{1,0} + S_{2,1} + 2S_{3,0}$$

4. **Add Round Key**: Each element of the 4 * 4 array, A, resulting from step 3, is XORed with the same element of array K encrypting the plaintext which results in the ciphertext as shown in Fig. 14.12.

Fig. 14.10 Result of shift process

$S_{0,0}$	$S_{0,1}$	$S_{0,2}$	$S_{0,3}$
$S_{1,1}$	$S_{1,2}$	$S_{1,3}$	$S_{1,0}$
$S_{2,2}$	$S_{2,3}$	$S_{2,0}$	$S_{2,1}$
$S_{3,3}$	$S_{3,0}$	$S_{3,1}$	$S_{3,2}$

$$
\begin{vmatrix} 2 & 3 & 1 & 1 \\ 1 & 2 & 3 & 1 \\ 1 & 1 & 2 & 3 \\ 3 & 1 & 1 & 2 \end{vmatrix} \times \begin{vmatrix} S_{0,0} & S_{0,1} & S_{0,2} & S_{0,3} \\ S_{1,1} & S_{1,2} & S_{1,3} & S_{1,0} \\ S_{2,2} & S_{2,3} & S_{2,0} & S_{2,1} \\ S_{3,3} & S_{3,0} & S_{3,1} & S_{3,2} \end{vmatrix} = \begin{vmatrix} A_{0,0} & A_{0,1} & A_{0,2} & A_{0,3} \\ A_{1,0} & A_{1,1} & A_{1,2} & A_{1,3} \\ A_{2,0} & A_{2,1} & A_{2,2} & A_{2,3} \\ A_{3,0} & A_{3,1} & A_{3,2} & A_{3,3} \end{vmatrix}
$$

Fig. 14.11 Mix column operation in AES

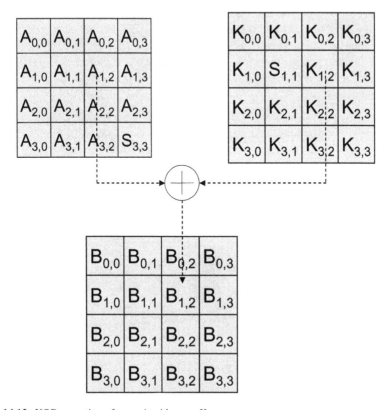

Fig. 14.12 XOR operation of array A with array K

5. The result of step 4, array B, becomes the new state for step 1 and this process is repeated 10 times. The encryption key is generated by key generator for each round

14.6   Asymmetric Cryptography

Public key cryptography, which is also called **asymmetric cryptography**, uses a different key for encryption and decryption. Public key cryptography uses two keys: a public key and a private key where the private key is kept secret. Figure 14.13a shows the plaintext encrypted by the public key and decrypted by the private key. Figure 14.13b shows the plaintext encrypted by the private key and decrypted by the public key.

Figure 14.14 shows that the key generator produces the public and private key. The key generator sends the public key to the transmitter. Then, the transmitter uses the public key and an encryption algorithm to produce the ciphertext. The transmitter then sends the ciphertext to the receiver and the receiver uses the private key and a decryption algorithm to decrypt the ciphertext.

The application of public key cryptography can be illustrated as follows: Assume a stockbroker serving 100 customers as shown in Fig. 14.15. The server generates a public key Kp for each customer, private key Ks, and a number N. When any customer connects to the server, the server passes the customer Kp and N. The customer uses these two numbers to encrypt information and transmit ciphertext to the broker server. The server then uses the private key Kp of the specific customer to decrypt the information.

Applications of Public Key Cryptography
Public key cryptography can be used for privacy, authentication, and non-repudiation. Some of the public key algorithms are as follows:

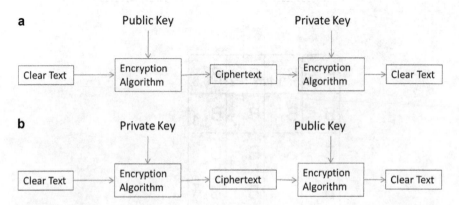

Fig. 14.13 (a) Public key encryption and private key decryption. (b) Private key encryption and public key encryption

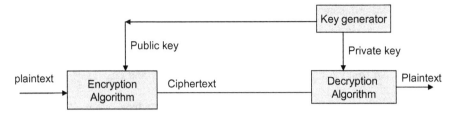

Fig. 14.14 Key generator

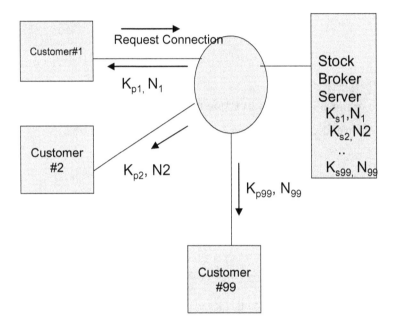

Fig. 14.15 Application of public key cryptography

A. **RSA**: RSA was named after three inventors: Ron Rivest, Adi Shamir, and Leonard Adleman, and it is a standard for public key cryptography algorithms.
B. **Diffie–Hellman Key Exchange**: The **Diffie–Hellman** key exchange enables two parties to establish a shared secret key over an insecure network.

14.7   RSA Algorithm

One of the most useful public key cryptography algorithms is the RSA, which is named after its inventors Ron Rivest, Adi Shamir, and Leonard Adleman (already stated above). RSA is based on number theory and works as follows:

1. Select two prime numbers p and q (prime number is a number divisible by 1 and itself only). The larger the number, the harder it is to break the RSA algorithm.
2. Let N = p * q, where N is the product of two prime numbers p and q.
3. Let Z = (p − 1) * (q − 1), where Z is the product of p − 1 and q − 1.
4. Select Kp (public key) such that Kp is less than N and has no common factors of Z.
5. Find Ks (secret key) such that Ks * Kp − 1 is divisible by Z or Ks = (1 + n * Z)/Kp; where n = 0, 1, 2, . . .
6. The transmitter uses Kp and N to encrypt message M using Eq. 14.1.

$$EM = M^{kp} \text{modulo N} \qquad\qquad (14.1)$$

where:

EM is ciphertext

The transmitter then transmits EM to the receiver.

7. The receiver decrypts EM using Ks and N according to Eq. 14.2.

$$M = EM^{Ks} \text{Modulo N} \qquad\qquad (14.2)$$

Example Find Ks and Kp assuming p = 3 and q = 5, also encrypt message M = 2 at the transmitted side then decrypt ciphertext at the receiving side (in this example, the prime numbers are selected for ease of understanding).

N = 3 * 5 = 15 and Z = (3 − 1) (5 − 1) = 8
Select a number that is less than N and is not a factor of Z (for this example, Kp = 3).
Ks = (1 + n * 8)/3 and n = 4, thus Ks = 11
Transmitter uses Kp to encrypt M = 3

EM = 3^3 modulo 15 = 12
The transmitter transmits cipher text of 12 to the receiver, then the receiver uses the secret key with N to decrypt the ciphertext of value 12.

M = EM^{KS} modulo N
or
M = 12^{11} modulo 15 = 3

14.8   Diffie–Hellman Key Exchange Algorithm

The Diffie–Hellman Algorithm is used by two parties to generate a symmetric key. The following steps describe Diffie–Hellman key generation.

1. Alice and Bob share two random numbers: G for the generator and P for a prime number.
2. Alice uses equation A = G^N Modulo P to generate A, where N is Alice's secret number.
3. Alice transmits A, P, and G to the Bob
4. Bob calculates B = G^M Modulo P, where M is Bob's secret number.
5. Bob transmits B to Alice.
6. Alice uses the following equation to calculate the symmetric key.

$$KA = B^N \text{ Modulo } P$$

7. Bob uses the following equation to calculate KB.

$$KB = A^M \text{ Modulo } P$$

8. As result the KA = KB.

Example

1. Alice selects G = 5, P = 7, and N = 3.
2. Alice calculates A = G^N Modulo P, A = 5^3 modulo 7, results A = 6.
3. Alice transmits A = 6, P = 7 and G = 5 to Bob.
4. Bob calculates B, B = G^M Modulo P, where M is a secret number, B = 5^5 modulo 7 = 3.
5. Bob transmits B = 3 to Alice.
6. Bob calculates KB = A^M Modulo P, KB = 6^5 modulo 7 = 6.
7. Alice calculates the secret key.
8. KA = B^N Modulo P, Ks = 3^3 modulo 7 = 6.
9. Result shows KA = KB.

14.9 Elliptic Curve Cryptography (ECC)

Elliptic curve cryptography (ECC) is used to implement symmetric key cryptography. ECC was invented by Victor Miller of IBM and Neil Koblitz of the University of Washington in 1985. The ECC offers a shorter key length with the same security provided by RSA, with the following advantages:

- **Shorter key lengths**: ECC uses 128 bits for encryption and 256 bits for digital signatures, with RSA needing 3072 bits to provide the same security as ECC.
- **Very fast key generation**: ECC generates symmetric keys faster than RSA.
- **Faster signature generation:** ECC generates digital signatures faster than RSA.
- **Faster encryption and decryption**: ECC implements faster encryption and decryption using ECDH than RSA.

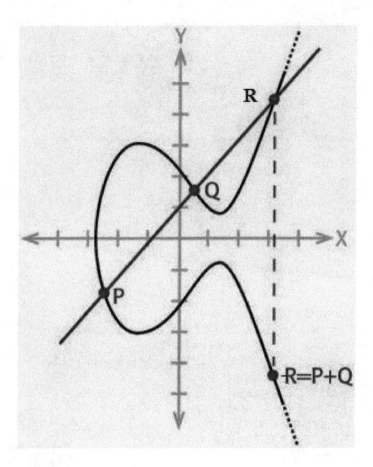

Fig. 14.16 Elliptic curve

Elliptic Curve Mathematics

An elliptic curve is a two-dimensional curve that is represented by the equation $y^2 = x^3 + ax + b$. The equation will not have repeated factors if $4a^3 + 27b^2 \neq 0$. For any given value of X on the curve, there is a Y value reflected over the x-axis. Figure 14.16 shows a general image of an elliptic curve and a non-vertical line that intersects the curve at three points.

A. **Finding points on the curve:** The are infinite points on an elliptic curve, and in order to limit the number of points available, elliptic curve cryptography works with a limited field of numbers as shown in Eq. 14.3:

$$y^2 \bmod P = \left(x^3 + ax + b\right) \bmod p, \text{ where the P is prime a number} \quad (14.3)$$

Example Find the points on the curve y^2 mod $11 = (x^3 + x + 3)$ Mod 11, where in the above equation $a = 1$, $b = 3$, and $P = 11$.

By using the equation $4a^3 + 27b^2 \neq 0$ to verify that $y^2 = (x^3 + x + 3)$ does not have repeated factors, therefore $4 * 1^3 + 27 * 3^2 \neq 0$.

Table 14.3 shows the values of X from 0 to 10 and the corresponding values for $(x^3 + x + 3)$ Mod 11. Also shown are the values of Y from 0 to 10 and the corresponding value of y^2 Mod 11.

By observing the results of Table 14.3:

When $(x^3 + x + 3)$ mod $11 = Y^2$ mod $11 = 3$, the results are points (0, 5) and (0, 6).

$(x^3 + x + 3)$ Mod $11 = Y^2$ Mod $11 = 5$ results in the points (1, 4), (1, 7), (4, 4), (4, 7), (0, 4), and (0, 7).

$(x^3 + x + 3)$ Mod $11 = Y^2$ Mod $11 = 0$ results in the point (3, 0).

$(x^3 + x + 3)$ Mod $11 = Y^2$ Mod $11 = 1$ results in the points (7, 1), (7, 10), (10, 10), and (10, 1).

$(x^3 + x + 3)$ Moe $11 = Y^2$ Mod $11 = 4$ results in the points (8, 2), (8, 9), (9, 2), and (9, 9).

Therefore, the points on the curve are: (0, 5), (0, 6), (1, 4), (1, 7), (3, 0), (4, 4), (4, 7), (5, 1), (5, 10), (6, 4), (6, 7), (7, 1), (7, 10), (9, 2), (9, 9), (10, 1), (10, 10)

B. **Adding points on the elliptic curve:** The elliptic curve is symmetrical about the x-axis. When given any point P, point $-P$ has to be the point opposite to P.

If P and Q are two points on the curve, then they can uniquely describe a third point, $P + Q$. by drawing the line that intersects P and Q it will intersect the curve at point R, where $P + Q$ is $-R$, or the point opposite of R, as shown in Fig. 14.17. The following properties apply for adding two points:

$P + Q = R, \quad Q + P = R, \quad P + P = 2P$
If $P = P(x, y)$, then $-P = P(x, -y)$
$P - P = 0$.

Finding Point R

Assume points P, Q, and R are represented by $P = (x_p, y_p)$, $Q = (x_q, y_q)$, and $R = (x_r, y_r)$.

Table 14.3 $(x^3 + x + 3)$ mod 11 and Y^2 mod 11

x	F1 = $(x^3 + x + 3)$ mod 11	y	F2 = y^2 mod 11
0	3	0	0
1	5	1	1
2	2	2	4
3	0	3	9
4	5	4	5
5	1	5	3
0	5	6	3
7	1	7	5
8	4	8	9
9	4	9	4
10	1	10	1

Fig. 14.17 Adding point P
to itself

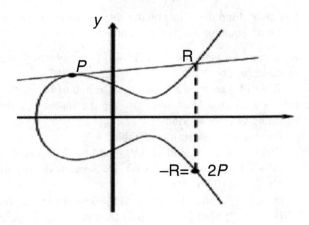

The slope of the line PQ is defined by:

$$m = \frac{y_q - y_p}{x_q - x_p}$$

and the point R is calculated by:

$$x_r = \left(m^2 - x_q - x_p\right) \bmod \ p$$

$$y_r = \left(m\left(x_r - x_q\right) - y_r\right) \bmod \ p$$

Example Adding points P(1, 4) and Q(7, 10)

$$m = \frac{10-4}{7-1} = \frac{6}{6} = 1$$

Xr = (1 − 1 − 7) = −7 mod 11 = 4
Yr = (1(−7 − 7) − 4) = −18 mod 11 = 4
R (4, 4), which is listed on the points discovered in the previous section.
Adding a point to itself (P + P = 2P)
    To add a point P to itself, a tangent line to the curve is drawn at the point P as shown in Fig. 14.17. If YP is not 0, then the tangent line intersects the elliptic curve at exactly one other point, −R, where −R is reflection of R on the x-axis.

$$P + P = 2P$$

The slope of the line that passes through the ECC curve is defined by:

$$m = \frac{3x_p^{\,2} + a}{2y_p}$$

The point Xr and Yr are calculated as follows:

$$x_r = \left(m^2 - 2x_p\right) \bmod P$$

$$y_r = \left(m\left(x_p - x_r\right) - y_p\right) \bmod P$$

Example Find the P + P = 2P for P(1, 4):

$$m = \frac{3*(1)^2 + 1}{2*4} = \frac{1}{2}$$

$$x_r = \left(1/2^2 - 2*1\right)\bmod 11 = -7/4 \ \bmod 11 = \frac{-7 \ \bmod 11}{4 \ \bmod 11} = 4/4 = 1$$

$$Yr = \left(\frac{1}{2}*(1-1)-4\right) \ \bmod 11 = -4 \ \bmod 11 = 7$$

Therefore, 2p = (1, 7).

Mr. Nicholas Brenckle from Southern Connecticut State University has developed a program that can find the points on the elliptic curve and the sum the two points, found at: http://ecdh.southernct.edu

Table 14.4 shows the resulting points and sums of those points on an elliptic curve $y^2 \bmod 11 = (x^3 + x + 3) \bmod 11$ using http://ecdh.southernct.edu

Table 14.4 Points and sums of points

+	(0,5)	(0,6)	(1,4)	(1,7)	(3,0)	(4,4)	(4,7)	(5,1)	(5,10)	(6,4)	(6,7)	(7,1)	(7,10)	(9,2)	(9,9)	(10,1)	(10,10)	∞
(0,5)	(1,7)	∞	(0,6)	(3,0)	(1,4)	(5,10)	(10,1)	(4,7)	(7,10)	(9,2)	(10,10)	(5,1)	(9,9)	(7,1)	(6,7)	(6,4)	(4,4)	(0,5)
(0,6)	∞	(1,4)	(3,0)	(0,5)	(1,7)	(10,10)	(5,1)	(7,1)	(4,4)	(10,1)	(9,9)	(9,2)	(5,10)	(6,4)	(7,10)	(4,7)	(6,7)	(0,6)
(1,4)	(0,6)	(3,0)	(1,7)	∞	(0,5)	(6,7)	(7,1)	(9,2)	(10,10)	(4,7)	(7,10)	(6,4)	(4,4)	(10,1)	(5,10)	(5,1)	(9,9)	(1,4)
(1,7)	(3,0)	(0,5)	∞	(1,4)	(0,6)	(7,10)	(6,4)	(10,1)	(9,9)	(7,1)	(4,4)	(4,7)	(6,7)	(5,1)	(10,10)	(9,2)	(5,10)	(1,7)
(3,0)	(1,4)	(1,7)	(0,5)	(0,6)	∞	(9,9)	(9,2)	(6,4)	(6,7)	(5,1)	(5,10)	(10,1)	(10,10)	(4,7)	(4,4)	(7,1)	(7,10)	(3,0)
(4,4)	(5,10)	(10,10)	(6,7)	(7,10)	(9,9)	(7,1)	∞	(0,6)	(5,1)	(1,7)	(6,4)	(1,4)	(4,7)	(3,0)	(10,1)	(0,5)	(9,2)	(4,4)
(4,7)	(10,1)	(5,1)	(7,1)	(6,4)	(9,2)	∞	(7,10)	(5,10)	(0,5)	(6,7)	(1,4)	(4,4)	(1,7)	(10,10)	(3,0)	(9,9)	(0,6)	(4,7)
(5,1)	(4,7)	(7,1)	(9,2)	(10,1)	(6,4)	(0,6)	(5,10)	(4,4)	∞	(9,9)	(3,0)	(10,10)	(0,5)	(6,7)	(1,7)	(7,10)	(1,4)	(5,1)
(5,10)	(7,10)	(4,4)	(10,10)	(9,9)	(6,7)	(5,1)	(0,5)	∞	(4,7)	(3,0)	(9,2)	(0,6)	(10,1)	(1,4)	(6,4)	(1,7)	(7,1)	(5,10)
(6,4)	(9,2)	(10,1)	(4,7)	(7,1)	(5,1)	(1,7)	(6,7)	(9,9)	(3,0)	(4,4)	∞	(7,10)	(1,4)	(5,10)	(0,6)	(10,10)	(0,5)	(6,4)
(6,7)	(10,10)	(9,9)	(7,10)	(4,4)	(5,10)	(6,4)	(1,4)	(3,0)	(9,2)	∞	(4,7)	(1,7)	(7,1)	(0,5)	(5,1)	(0,6)	(10,1)	(6,7)
(7,1)	(5,1)	(9,2)	(6,4)	(4,7)	(10,1)	(1,4)	(4,4)	(10,10)	(0,6)	(7,10)	(1,7)	(6,7)	∞	(9,9)	(0,5)	(5,10)	(3,0)	(7,1)
(7,10)	(9,9)	(5,10)	(4,4)	(6,7)	(10,10)	(4,7)	(1,7)	(0,5)	(10,1)	(1,4)	(7,1)	∞	(6,4)	(0,6)	(9,2)	(3,0)	(5,1)	(7,10)
(9,2)	(7,1)	(6,4)	(10,1)	(5,1)	(4,7)	(3,0)	(10,10)	(6,7)	(1,4)	(5,10)	(0,5)	(9,9)	(0,6)	(7,10)	∞	(4,4)	(1,7)	(9,2)
(9,9)	(6,7)	(7,10)	(5,10)	(10,10)	(4,4)	(10,1)	(3,0)	(1,7)	(6,4)	(0,6)	(5,1)	(0,5)	(9,2)	∞	(7,1)	(1,4)	(4,7)	(9,9)
(10,1)	(6,4)	(4,7)	(5,1)	(9,2)	(7,1)	(0,5)	(9,9)	(7,10)	(1,7)	(10,10)	(0,6)	(5,10)	(3,0)	(4,4)	(1,4)	(6,7)	∞	(10,1)
(10,10)	(4,4)	(6,7)	(9,9)	(5,10)	(7,10)	(9,2)	(0,6)	(1,4)	(7,1)	(0,5)	(10,1)	(3,0)	(5,1)	(1,7)	(4,7)	∞	(6,4)	(10,10)
∞	(0,5)	(0,6)	(1,4)	(1,7)	(3,0)	(4,4)	(4,7)	(5,1)	(5,10)	(6,4)	(6,7)	(7,1)	(7,10)	(9,2)	(9,9)	(10,1)	(10,10)	∞

Elliptical Curve Diffie–Hellman (ECDH) Key Exchange

With the Diffie–Hellman key exchange, a public key cryptosystem can be developed. This is known as "ECDH," or the elliptic curve Diffie–Hellman method.

After two parties, Alice and Bob, select common curve parameters (a, b, P) and a common point along the curve called G, they can then use the following steps to develop a symmetric key.

Alice and Bob select a point on the curve as the generator (G), and they calculate the cofactor (h) as follows:

They use the G to find 2G, 3G, 4G, ... nG, where nG reaches infinity.

Example: If Alice and Bob select G(4, 4) and use the table's addition to find:

2G = G + G = (7, 1), 3G = 2G + G = (1, 4), 4G = 2G +2G = (6, 7), 5G = (6, 4), 6G = (1, 7), 7G = (7, 10), 8G = (4, 7), and 9G = 8G + G = infinity, then n = 8.

The following represents G,2G.........up to 8G

(4, 4), (7, 1), (1, 4), (6, 7), (6, 4), (1, 7), (7, 10), (4, 7)

h = the number of points on the curve / n = 17 / 8 = 2.1

The smaller h shows a good selection of G and results in better security.

The following steps describe the generation of a public key using ECDH:

- Alice and Bob select a secret random number between 1 and n. Alice's secret number is N, and Bob's secret number is M.
- Alice calculates the public key (AP) = N * G, and sends it to Bob.
- Bob calculates the public key (BP) = M * G, and sends it to Alice.
- Alice uses Bob's public key (BP) to generate the symmetric key (SK): SK = N * M *G.
- Bob uses Alice's public key (AP) to generate the symmetric key (SK): SK = M * N *G.

Example: Find the secret keys for Alice and Bob, assuming Alice's random number N = 3 and Bob's random number M = 5, with a generator G = (1, 4).

Alice's public key AP = N * G = 3G = 3 * (10, 10), which is (10, 10) + (10, 10) + (10, 10). Table x.1 shows that (10, 10) + (10, 10) = (6, 4), and then (6, 4) + (10, 10) = (0, 5), so AP = (0, 5).

Bob's public key BP = M * G = 5 * (10, 10), or (0, 5) + (10, 10) + (10, 10). Again, using Table 14.1, (0, 5) + (10, 10) = (4, 4), and (4, 4) + (10, 10) = (9, 2), so BP = (9, 2).

Alice and Bob exchange their public keys.

Alice uses Bob's public key to generate their secret key (AS).

Alice's secret key AS = N * (BP) = 3 * (9, 2) = (0, 6).

Bob's secret key BS = M * AP = 5 * (0, 5) = (0, 6).

Encryption and Decryption using ECDH

Alice selects a point on the curve which represents their message and encrypts the point. Then, the ciphertext is transmitted to Bob.

Assume that Alice's message is S = (3, 0), Alice uses Eq. 14.4 for the encryption of S:

$$C = AP, (S + N^*BP) = Y1, Y2 \qquad (14.4)$$

where C is cypher text, AP is Alice's public key, N is Alice's random number, and BP is Bob's public key. Using Table 14.4 to evaluate, the results are as follows:

$$C = (0,5), ((3,0) + (0,6)) = (0,5), (1,7)$$

Bob uses Eq. 14.3 to decrypt C:

$$S = Y2 - Nb^*Y1 \qquad\qquad (14.5)$$

S = (1, 7) − 5(0, 5) = (1, 7) − (0, 6) = (1, 7) + (0, −6)
−6 mod 11 = 5
M = (1, 7) + (0, 5) = (3, 0)

14.10   Hash Value or Message Digest

A hash function generates a hash value (message digest) from a given message as shown in Fig. 14.18, and it is used for data integrity and password protection. Hash values have the following characteristics:

1. A hash function must be collision resistant which means it is hard to find two messages that will generate the same message digest.
2. Changing any bit of message will totally change the message digest.
3. It is impossible to generate the original message from the message digest.

Applications of hash function include hashing passwords and verifying message integrity

Hashing Passwords If a server holds encrypted passwords and a hacker accesses the password file and has the encryption key, then the passwords will be open to decryption. By hashing the passwords, the attacker will not be able to recover the passwords. When a user enters their password, the server will use the hash algorithm and generate the hash value of the password. If the hash value is the same as the stored hash value, then the password is correct.

The following are some of the basic hash functions:

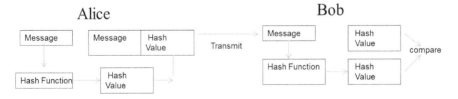

Fig. 14.18 Data integrity using message digest

Frame Check Sequence (FCS) as a Hash Value: Host B generates the FCS of the information, encrypts the FCS, and sends it to Host A. If one alters the document, the new FCS will differ from the one generated by Host B.

Checksum (generating checksum from a message): Information can be broken into blocks of characters and arranged in each column, the message $6,578,100 can be broken into 4 ASCII characters per block as shown below, and the checksum of each column can be calculated

$657	44	36	35	37
8100	38	31	30	30
Checksum	7C	67	65	67

If the message changes to $5,578,200, it will generate the same checksum; therefore, the checksum method is not a strong message digest. The simplest form of a hash function is to break the message into m blocks of n bits as shown below.

Block0	b_{01}	b_{02}	b_{03}	...	b_{0n}
Block1	b_{11}	b_{12}	b_{13}	...	b_{1n}
...	...	...	...	...	...
Blockm	b_{m1}	b_{m2}	b_{m3}	...	b_{mn}

Hash bits are represented by:

$$H = H_1 H_2 H_3 \ldots Hn$$

where:

$$H_1 = b_{01} \text{XORed } b_{11} \text{XORed } b_{21} \text{ XORed} \ldots \text{XORed bml}$$

Or in general:

$$Hi = b_{i1} \text{XORed } b_{i2} \text{XORed } b_{i3} \text{XORed } b_{i4} \text{XORed} \ldots \text{XORed } b_{in}$$

Example 14.3 Find the hash code for the word "WELCOME."

W	1010111
E	1000101
L	1001100
C	1000011
O	1001111
M	1001101
E	1000101
Hash code	1001001

Some of the most popular message digests are as follows:

A. **MD4**: MD4 is a message digest type 4 that generates a 128-bit hash value or message digest.
B. **MD5:** MD5 is an improved version of MD4 that generates a 128-bit hash value
C. **SHA-1 (Secure Hash Algorithm 1)**: SHA-1 generates a 160-bit hash
D. **SHA-256**

Data Integrity
A hash function is used for verifying data integrity. Alice generates a hash value from a message and attaches it to the message and transmits it to Bob as shown in Fig. 14.18. Bob takes the message and generates a hash value using the same hashing function used by Alice. If the hash value generated by the Bob is the same as the hash value generated by Alice, then nobody tampered with the message. Figure 14.18 shows the application of hash value for data integrity.

What happens if another person generates a message and message digest instead of Alice and transits it to Bob? How will Bob know if this message is from Alice or not? The Message Authentication Code (MAC) is used for authentication of sender messages and data integrity.

14.11   Message Authentication Code (MAC)

MAC is used for message integrity and authentication of the sender. To generate a MAC both Alice and Bob must share a secret key. The secret key is used for authentication, and the following steps describe the message authentication process. Figure 14.19 shows the Message Authentication Code generation and usage.

1. Alice generates a hash value from the message and encrypts it with the shared key. This is called the MAC.
2. Alice attaches the MAC to the message and transmits it to Bob.
3. Bob generates the hash value of the message and decrypts the MAC to generate a Hash value.
4. If both hashes are equal, there was no message tampering.

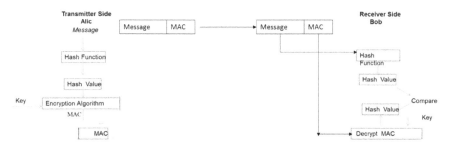

Fig. 14.19 Data integrity and authentication

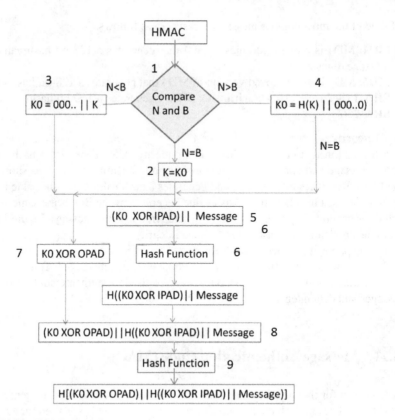

Fig. 14.20 HMAC generation

Keyed-Hash Message Authentication (HMAC)

HAMC is a message authentic code where the sender and the receiver share a secret key. The application of HMAC is to authenticate the source of a message and the integrity of a message. A HMAC is generated using a secret key and message. Figure 14.20 shows HMAC generation where:

K: Secret key.

N: Size of secret key in bytes.

B: Size of the message in bytes.

K0: If the message is the same size as K (secret key), then K = K0.

IPAD (Inner Pad): IPAD = 0x3636 ... the 36 will repeat in order to become the same sizeofsecretkey,ifthesecretkey(K)is10bytesthenIPAD=0x3636363636363636363636 (each byte can be represented by two digits in hex).

OPAD (Outer Pad): OPAD = 0x5C 5C......, the 5C repeated until the size of OPAD become the same size as K.

‖: Means concatenation, concatenation of 34 and 76 is 3476.

L: Number of bytes generated by Hash.

The following steps describe HAMC generation:

1. Compare the N (number of bytes in secret key) with B (number of bytes in message).
2. If they are equal, then K0 = K. Jump to step 4.
3. If N is less than B, then add zeroes to the left of the K in order N = B. Then K=K0.
4. If N is greater than B, then find the Hash value of K (H(K)) which is L bytes, then left pad the result with L-B bytes of zeros, resulting in K=B.
5. K0 XOR with IPAD concatenated with message results (K0 XOR IPAD) ‖ Message.
   IPAD = 0x36 and the 36 repeated in order that Size of IPAD = N.
   Hash the result of step 5: **H ((K0 XOR IPAD) ‖ Message.**
6. OPAD = 0x5C and 5C repeated in order the size of OPAD become N bytes then K0 XORed with OPAD K0 XOR OPAD.
7. **K0 XOR OPAD** concatenated with **H((K0 XOR IPAD)‖ Message** results in **(K0 XOR OPAD)‖H((K0 XOR IPAD)‖ Message).**
8. The result of step 8 passes through Hash function results H[**(K0 XOR OPAD)‖H((K0 XOR IPAD)‖ Message)]** which is the HAMC.

14.12   Digital Signature

In business transactions, we sign various documents such as checks and contracts. The reason we sign documents is to indicate our understanding and acceptance of the contents of the document. The objective of a digital signature is to sign an electronic document rather than a paper document. Public cryptography is one of the methods used for digital signatures. An example of this is shown in Fig. 14.21. In Fig. 14.21, Alice sends a document to Bob. Bob encrypts the document with its private key and sends it back to Alice. Alice stores the encrypted document in case Bob denies the signature. Alice decrypts the document for her use.

Fig. 14.21 Digital signature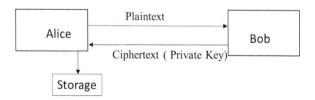

Transmitter side   Receiver Side
Alice   Bob

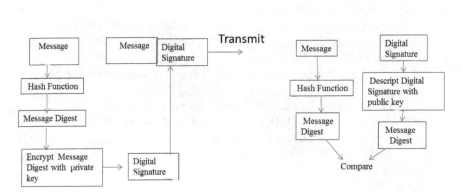

Fig. 14.22 Digital signature using private key

The disadvantage of the above method is that the entire document must be encrypted and stored on the receiver side thus requiring a large amount of memory. Another method is when one signs a message digest (i.e., a summary of the document contents) rather than the entire document. A message digest is a summary of the message such as a frame check sequence. A hash value encrypted by private key is called digital signature as shown in Fig. 14.22.

14.13   Kerberos

Kerberos is an authentication system that was developed at the Massachusetts Institute of Technology (MIT). It is used for the authentication of two systems over an unsecured network. The components of Kerberos are shown in Fig. 14.23. For Host A to access Server B, the following steps should take place:

1. User A logs into the Kerberos server (**KS**) and requests to access Server B.
2. The KS prompt the user A for a password, a user ID, and a server ID.
3. KS checks the user's password and ID and also checks if user A has permission to access Server B.
4. If user A passes step 3, then KS creates a ticket that contains a user ID and a server ID. KS encrypts the ticket and transmits it to user A and server B. Only server B can encrypt the ticket.

Fig. 14.23 Kerberos
authentication system

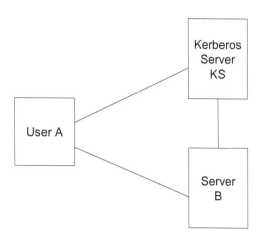

5. User A sends the ticket to server B; server B encrypts the ticket and compares it with the ticket that was sent by KS. If both tickets are the same, then user A is allowed to access server B.

Summary

- A direct attack is when the attacker breaks the security of a system.
- An indirect attack is when the attacker gets information and data (such as a name, address, social security, or credit card number).
- The elements of network security are confidentiality, authentication, integrity, and non-repudiation.
- When plaintext is encrypted by an encryption algorithm, it is called a ciphertext.
- The algorithm that is used for encryption and decryption is called a cipher.
- Cryptography is the science of keeping information secret.
- Modern cryptography is classified into two types: symmetric cryptography (private key) and asymmetric cryptography (public key).
- In symmetric cryptography, the transmitter and the receiver use the same key for encryption and decryption.
- In asymmetric cryptography (public key), the transmitter and the receiver use different keys for encryption and decryption.
- The Data Encryption Algorithm is an example of symmetric cryptography.
- The elliptic curve equation is $y^2 = x^3 + ax + b$.
- The RSA algorithm is an example of public key cryptography.
- A digital signature is a process that is used by a client to sign an electronic document.

Key Terms

Authentication	Firewall
Authenticator	Hash function
Block cipher	Hash value
Certificates	Indirect attack
Ciphertext	Integrity
Cryptography	Kerberos
Data Encryption Standard	Mutual authentication
Decryption	Public key cryptography
Digital signature	RSA algorithm
Direct attacks	Stream cipher

Review Questions

Multiple Choice Questions

1. _____ provides privacy and protects information from an attacker.

   (a) Authentication
   (b) Confidentiality
   (c) Integrity
   (d) Non-repudiation

2. Plaintext encrypted is called a/an _____

   (a) Cipher
   (b) Encryption
   (c) Ciphertext
   (d) Hash value

3. When a transmitter and a receiver use the same key, the key is called a
   _____

   (a) Public key
   (b) Private key
   (c) Secret key
   (d) Hash key

4. The science of encryption and decryption is called _____

   (a) Authentication
   (b) Cipher
   (c) Cryptography
   (d) Hash value

5. The algorithm is used for encryption and decryption is called a _____

    (a) Cleartext
    (b) Cipher
    (c) Ciphertext
    (d) Digital signature

6. Stream ciphers encrypt information _____

    (a) One bit at a time
    (b) One byte at a time
    (c) One block at time
    (d) All at the same time

7. Which of following algorithms use a private key?

    (a) DES
    (b) RAS
    (c) Hash function
    (d) Digital signature

8. Computer A uses a public key for encrypting its information and sends it to station B. What does Computer B Use for decryption?

    (a) Computer B uses its public key for decryption
    (b) Computer B uses its private key for decryption
    (c) Computer B uses Computer A's public key for decryption
    (d) Computer B randomly chooses a key for decryption

9. The objective of a digital signature is _____

    (a) To verify the identity of a user or a client
    (b) To provide privacy for a document
    (c) To inform the acceptance of the document by user
    (d) For the user to copy the document

10. A hash function is used for _____

    (a) Digital signature
    (b) Digital certificates
    (c) Encryption
    (d) Authentication

11. Keberos is used for:

    (a) Encryption of a document
    (b) Digital signatures
    (c) Digital certificates
    (d) Authentication purposes
    (e) None of the above

Short Answer Questions

1. List the types of attacks and explain each type.
2. List the elements of network security and explain the function of each.
3. List some of the common attacks in networking.
4. What is the definition of cryptography?
5. Show the cryptography model.
6. What is a cipher?
7. List the classes of cryptography.
8. Explain symmetric cryptography.
9. Explain asymmetric cryptography.
10. Distinguish between stream ciphers and block ciphers
11. Name an encryption algorithm that uses a stream cipher.
12. List three encryption algorithms that use block ciphers.
13. What does DES stand for?
14. What does AES stand for?
15. What is 3DES?
16. Name an algorithm which generates asymmetric keys.
17. What are the applications of public key cryptography?
18. What are the characteristics of a hash value?
19. What is the function of a hash function?
20. What is the equation for elliptic curve?
21. What is a condition for the elliptic curve in order to not have repeated factors?
22. What are the advantages of ECC?
23. What are some popular hash functions?
24. Explain digital signatures.
25. Explain the function of Kerberos.
26. Find the public key, the private key, and N using the RSA (assume the two prime numbers $p = 7$ and $q = 11$).
27. Encrypt the message $M = 5$

    (a) Using the public key in Problem 20.
    (b) Decrypt the encrypted message using the private key in Problem 20,

28. What are the applications of MAC?
29. How is MAC generated?

Chapter 15
Network Security

Objectives

After completing this chapter, you should be able to:

- Explain the application of the Secure Socket Layer (SSL).
- Describe the components of SSL/TLS.
- Understand VPNs and how they operate.
- Explain the application of EAP.
- Explain the application of the SSH protocol.
- Discuss different applications of a firewall.
- Explain the purpose of certificates.
- Distinguish between WPA, WPA2, and WPA3.

Introduction

The transfer of information across networks and the Internet is increasing exponentially due to e-commerce and business transactions. Network security plays an important role in successful e-commerce. Currently, many people are accessing their bank accounts, buying and selling stocks, and paying bills over the Internet. People using these services need to have their transactions be secure. This means that the information transmitted should not be able to be accessed or modified by anyone other than the authorized user. Network security is implemented to protect information in transit and to protect a system from an attack.

15.1   Secure Socket Layer Protocol (SSL)

The Secure Socket Layer protocol (SSL) was developed by the Netscape Communication Corporation in 1995 for secure communication between a web browser and a web server. Version 3 is the latest version of SSL. The IETF modified SSL v3 and called it Transport Layer Security (TLS), which is what is typically seen

© The Author(s), under exclusive license to Springer Nature Switzerland AG 2024
A. Elahi, A. Cushman, *Computer Networks*,
https://doi.org/10.1007/978-3-031-42018-4_15

Fig. 15.1 SSL/TLS
protocol architecture

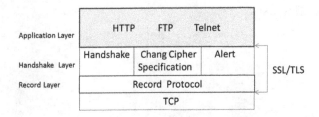

in use today. However, even when TLS is in use, the term "SSL" tends to be used interchangeably. The main difference between TLS and SSL is that TLS uses Keyed-Hash Message Authentication Code (HMAC), and SSL uses MAC. SSL/TLS provides data encryption, server authentication, message integrity, and client authentication (optional) for both ends.

The SSL/TLS protocol is located between the TCP protocol layer and the application protocol layer as shown in Fig. 15.1. HTTPS is Hypertext Transfer Protocol over Secure Socket Layer, which replaces HTTP. HTTPS uses port 443, where HTTP uses port 80 for interaction with TCP, as shown in Fig. 15.2. The SSL/TLS protocol consists of the following components:

A. The Handshake Layer consists of Handshake, Change Cipher Specification, and Alert Protocols
B. The Record Layer consists of Record Protocol

SSL Handshake Protocol
The SSL handshake protocol performs a series of message exchanges between an SSL client and SSL server to authenticate each other and negotiate the encryption algorithm established via a secure connection. SSL/TLS can use symmetric cryptography or public key cryptography and is decided based on the cryptography algorithm negotiation.

Figure 15.3 shows Handshake packets between a client and a server. The following steps describe the SSL/TLS handshake packets:

1. The client makes a connection to a server using the standard TCP handshake.
2. The client transmits a **Hello Packet** to the server as shown in Fig. 15.4. This Hello Packet contains the following information:

    (a) The SSL/TLS version.
    (b) **Random Number**: This random number, along with a random number and a third value referred to as the premaster key, is used to generate a master secret key from which the encryption key will be generated.
    (c) **Session ID**: This is used if the session becomes disconnected, alleviating the need for the client and server to perform another handshake.
    (d) **Cipher Suites**: This is a list of the client's supported cryptographic algorithms, message digest algorithms, and key exchange algorithms, which may be used for SSL handshaking. The list may contain up to 31 cipher suites.
    (e) **Compression method**: This is optional.

Fig. 15.2 HTTPS
architecture

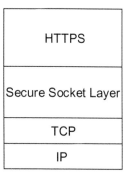

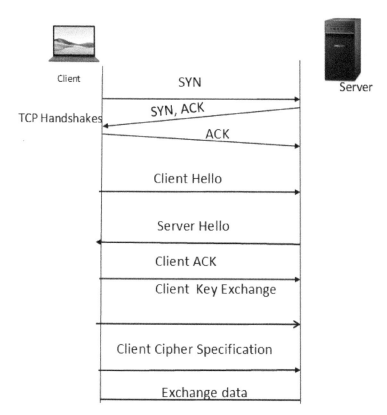

Fig. 15.3 SSL/TLS handshake packets

3. The server acknowledges the client's **Hello Packet**.
4. The server transmits its **Hello packet** to the client, as shown in Fig. 15.5, with the following information:

   (a) **Random Number**: Just like the client, the server also generates and sends over its own random number for later use.

```
⚟ 32 9.500495 10.72.8.18 10.64.24.200 SSLv3 Client Hello
⊞ Frame 32: 158 bytes on wire (1264 bits), 158 bytes captured (1264 bits)
⊞ Ethernet II, Src: Dell_22:b9:3c (00:21:70:22:b9:3c), Dst: Cisco_d7:28:00 (00:1c:f9:d7:28:00)
⊞ Internet Protocol, Src: 10.72.8.18 (10.72.8.18), Dst: 10.64.24.200 (10.64.24.200)
⊞ Transmission Control Protocol, Src Port: necp (3262), Dst Port: https (443), Seq: 1, Ack: 1, Len: 104
⊟ Secure Socket Layer
  ⊟ SSLv3 Record Layer: Handshake Protocol: Client Hello
      Content Type: Handshake (22)
      Version: SSL 3.0 (0x0300)
      Length: 99
    ⊟ Handshake Protocol: Client Hello
        Handshake Type: Client Hello (1)
        Length: 95
        Version: SSL 3.0 (0x0300)
      ⊟ Random
          gmt_unix_time: Jan 20, 2012 10:01:07.000000000 Eastern Standard Time
          random_bytes: 4bbea0846502909b182c65f1c7434da6ecc827886f164f58...
        Session ID Length: 32
        Session ID: b42500005b8dfffb880d7f8da91501572beae612b2daddf8...
        Cipher Suites Length: 24
      ⊟ Cipher Suites (12 suites)
          Cipher Suite: TLS_RSA_WITH_RC4_128_MD5 (0x0004)
          Cipher Suite: TLS_RSA_WITH_RC4_128_SHA (0x0005)
          Cipher Suite: TLS_RSA_WITH_3DES_EDE_CBC_SHA (0x000a)
          Cipher Suite: TLS_RSA_WITH_DES_CBC_SHA (0x0009)
          Cipher Suite: TLS_RSA_EXPORT1024_WITH_RC4_56_SHA (0x0064)
          Cipher Suite: TLS_RSA_EXPORT1024_WITH_DES_CBC_SHA (0x0062)
          Cipher Suite: TLS_RSA_EXPORT_WITH_RC4_40_MD5 (0x0003)
          Cipher Suite: TLS_RSA_EXPORT_WITH_RC2_CBC_40_MD5 (0x0006)
          Cipher Suite: TLS_DHE_DSS_WITH_3DES_EDE_CBC_SHA (0x0013)
          Cipher Suite: TLS_DHE_DSS_WITH_DES_CBC_SHA (0x0012)
          Cipher Suite: TLS_DHE_DSS_EXPORT1024_WITH_DES_CBC_SHA (0x0063)
          Cipher Suite: Unknown (0x00ff)
        Compression Methods Length: 1
      ⊟ Compression Methods (1 method)
          Compression Method: null (0)
```

Fig. 15.4   Hello Packet

```
⚟ 34 9.500928 10.64.24.200 10.72.8.18 SSLv3 Server Hello, Certificate, Server Hello Done
⊞ Frame 34: 1221 bytes on wire (9768 bits), 1221 bytes captured (9768 bits)
⊞ Ethernet II, Src: Cisco_d7:28:00 (00:1c:f9:d7:28:00), Dst: Dell_22:b9:3c (00:21:70:22:b9:3c)
⊞ Internet Protocol, Src: 10.64.24.200 (10.64.24.200), Dst: 10.72.8.18 (10.72.8.18)
⊞ Transmission Control Protocol, Src Port: https (443), Dst Port: necp (3262), Seq: 1461, Ack: 105, Len: 1167
⊞ [Reassembled TCP Segments (2627 bytes): #33(1460), #34(1167)]
⊟ Secure Socket Layer
  ⊟ SSLv3 Record Layer: Handshake Protocol: Multiple Handshake Messages
      Content Type: Handshake (22)
      Version: SSL 3.0 (0x0300)
      Length: 2622
    ⊟ Handshake Protocol: Server Hello
        Handshake Type: Server Hello (2)
        Length: 77
        Version: SSL 3.0 (0x0300)
      ⊟ Random
          gmt_unix_time: Jan 20, 2012 10:01:07.000000000 Eastern Standard Time
          random_bytes: 1fc0946eab95bdd871b1fc0fc08130c58c1179f85918825d...
        Session ID Length: 32
        Session ID: 6d010000f7f4badd3f3c3c05996cbdc104c0b528873d8e1a...
        Cipher Suite: TLS_RSA_WITH_RC4_128_MD5 (0x0004)
        Compression Method: null (0)
        Extensions Length: 5
      ⊞ Extension: renegotiation_info
    ⊞ Handshake Protocol: Certificate
    ⊞ Handshake Protocol: Server Hello Done
```

Fig. 15.5   Server Hello Packet

    (b) **Session ID.**

    (c) **Security Suite**: The server selects the first security suite from the list that
was transmitted by the client. If the server supports the security suite, then it
is used, otherwise the server selects the next one.

```
Frame 36: 386 bytes on wire (3088 bits), 386 bytes captured (3088 bits)
Ethernet II, Src: Dell_22:b9:3c (00:21:70:22:b9:3c), Dst: Cisco_d7:28:00
Internet Protocol, Src: 10.72.8.18 (10.72.8.18), Dst: 10.64.24.200 (10.6
Transmission Control Protocol, Src Port: necp (3262), Dst Port: https (4
Secure Socket Layer
  SSLv3 Record Layer: Handshake Protocol: Client Key Exchange
      Content Type: Handshake (22)
      Version: SSL 3.0 (0x0300)
      Length: 260
    Handshake Protocol: Client Key Exchange
        Handshake Type: Client Key Exchange (16)
        Length: 256
  SSLv3 Record Layer: Change Cipher Spec Protocol: Change Cipher Spec
      Content Type: Change Cipher Spec (20)
      Version: SSL 3.0 (0x0300)
      Length: 1
      Change Cipher Spec Message
  SSLv3 Record Layer: Handshake Protocol: Encrypted Handshake Message
      Content Type: Handshake (22)
      Version: SSL 3.0 (0x0300)
      Length: 56
      Handshake Protocol: Encrypted Handshake Message
```

Fig. 15.6 Client Key Exchange Packet

   (d) **Server Certificate**: The server provides the client with a digital certificate,
       which establishes the server's identity, contains the server's public key, and
       is digitally signed in a chain of trust. Digital certificates will be covered in
       detail in Sect. 15.6.
   (e) Handshake Done.

5. The client, now with the server's certificate, sends a Key Exchange Packet, as
   seen in Fig. 15.6, with the following information:

   (a) **Premaster Key**: Another random number is chosen by the client to be a
       premaster key, which is then encrypted using the public key provided in the
       server's certificate. Both the client and the server will use this premaster key,
       as well as the other random numbers generated by the client and server, to
       generate a symmetric session key. The server does so by decrypting the pre-
       master key that was just sent, which it should only be able to do if it has the
       corresponding private key for the public key provided in the certificate.
   (b) The client transmits a **Change Cipher Specification** packet to the server as
       shown in Fig. 15.7. This message notifies the server that all messages from
       now on will be encrypted using the key and algorithms negotiated. If the
       server is able to decrypt the premaster key, then the client and server should
       have identical session keys and should now be able to communicate securely.

Record Protocol The Record Protocol receives the data from the application
layer or TCP layer and performs following functions:

(a) Fragments the data into blocks or reassembles the data packet.
(b) Adds sequence numbers to each block.
(c) Compresses or decompresses the data packet using the compression algorithm
    negotiated in the handshake protocol.

```
Frame 37: 121 bytes on wire (968 bits), 121 bytes captured (968 bits)
Ethernet II, Src: Cisco_d7:28:00 (00:1c:f9:d7:28:00), Dst: Dell_22:b9:3c (00:21:7
Internet Protocol, Src: 10.64.24.200 (10.64.24.200), Dst: 10.72.8.18 (10.72.8.18)
Transmission Control Protocol, Src Port: https (443), Dst Port: necp (3262), Seq:
Secure Socket Layer
  SSLv3 Record Layer: Change Cipher Spec Protocol: Change Cipher Spec
    Content Type: Change Cipher Spec (20)
    Version: SSL 3.0 (0x0300)
    Length: 1
    Change Cipher Spec Message
  SSLv3 Record Layer: Handshake Protocol: Encrypted Handshake Message
    Content Type: Handshake (22)
    Version: SSL 3.0 (0x0300)
    Length: 56
    Handshake Protocol: Encrypted Handshake Message
```

Fig. 15.7 Change Cipher Specification

(d) Encrypts or decrypts the data.
(e) Applies an HMAC (or a MAC for SSL 3.0) to outgoing data.
(f) Computes the HMAC and verifies that it is identical to the value transmitted. This check ensures data integrity when a message is received.

Alert Protocol The Alert Protocol is used to notify both client and server about errors and session termination.

15.2   Virtual Private Network (VPN)

Large organizations and corporations have multiple sites in different locations and would often like to connect LANs together. One solution is to lease a data communication line and connect the LANs together thus creating a private network. A corporation with 100 offices in different locations throughout a country must lease 100 lines to connect its LANs together. This method is not cost-effective. In order to reduce the cost of leasing private lines, Virtual Private Networks (VPNs) can be employed over Internet.

Figure 15.8 shows two networks that are connected through the Internet using VPN devices. The function of a VPN device is to make a secure communication between two LANs by establishing a tunnel between its two endpoints. The disadvantage of using the Internet as a communication channel between corporations is poor security. Therefore, the VPN needs to provide security components such as authentication, confidentiality and data integrity.

Tunneling
Tunneling is the process of placing one packet inside another packet for transmission over a public network. In Fig. 15.8, it is assumed that both networks are running IP. If station A wants to send an IP packet to station B, it sends the packet to the VPN device A first. VPN device A then encapsulates the packet into an IP packet and transmits it over the Internet. When VPN device B receives the IP packet, it

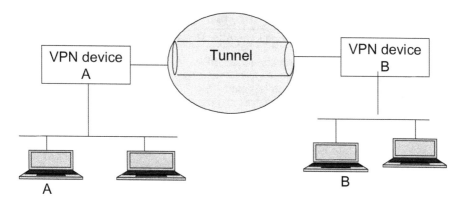

Fig. 15.8 Virtual Private Network (VPN)

discards the IP header and sends the IP packet to station B. Some of the tunneling protocols are listed below.

1. Point-to-Point Tunneling Protocol (PPTP): PPTP was developed by the PPTP forum. PPTP supports 40 and 128-bit encryption.
2. Layer-2 Forwarding (L2F): L2F was developed by CISCO and it uses any authentication method.
3. Layer-2 Tunneling Protocol (L2FT): L2FT was developed by CISCO and IETF. It is a combination of L2F and Point-to-Point Tunnel Protocol
4. IP security (IPsec) Protocol: The IP Security Protocol was developed by IETF.

15.3   IP Security Protocol (IPsec)

In order to protect information traveling via VPNs from attack, the IETF developed the IPsec protocol for integrity, authentication, and confidentiality of data over IP networks. The **IPsec** Protocol is an open standard and does not define any specific protocol for authentication and encryption. The IP Security Protocol is becoming increasingly more popular than the other protocols. IPsec is made of two protocols **Authentication Header Protocol (AH)** and **Encapsulation Security Protocol (ESP)**.

Authentication Header Protocol
The **Authentication Header Protocol (AH)** is used for identification of a user or application and it can operate in either tunneling mode or transport mode. AH does not provide any encryption. Figure 15.9 shows the AH packet format.
   The following describes the function of each field of authentication header:

- **Next Header**: Identifies the next header (such as TCP).
- **Payload Length**: Number of 32-bit words in AH.

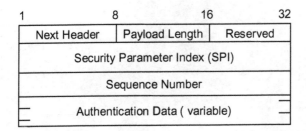

Fig. 15.9 AH packet format of IPsec

- **Security Parameter Index (SPI)**: The SPI informs the receiving station of the location of the security information such as the decryption key.
- **Sequence Number**: Every packet has a unique number. If an intruder copies a packet and resends it with the same sequence number, then the receiver discards the packet.
- **Authentication Data**: This is digital signature of the payload.

Figure 15.10 shows the AH location in an IPsec packet in transport mode. In the IP header, the protocol type is 50 thus indicating that the next header is an IPsec header. Figure 15.11 shows the location of AH in the IPsec packet using tunneling mode. In tunneling mode, the entire packet is encapsulated into another IP packet and AH includes the entire IP packet.

Encapsulation Security Packet (ESP)
The Encapsulation Security Packet performs encryption and authentication of data. Figure 15.12 shows an ESP format.

ESP can operate in tunneling mode or in transport mode. Figure 15.13 shows the ESP format using transport mode. In transport mode, only the TCP header and data fields are encrypted. The authentication data is a digital signature of the data, padding, TCP header, and ESP header. As shown in Fig. 15.13, the protocol type in IP header is 50 thus indicating that the next protocol is ESP. Figure 15.14 shows ESP using tunneling mode. In tunneling mode, the entire IP packet is encrypted and has the entire ESP header encapsulated into the IP packet for transmission.

15.4 Secure Shell (SSH)

SSH was developed in 1995 and is widely used in many Unix systems to overcome the weaknesses of Telnet and FTP. The SSH standards were developed by the IETF and then the Secure Shell was commercialized in 1998 by www.ssh.com where it was implemented for the Windows and UNIX systems. SSH version one (SSH1) had numerous weaknesses, but SSH2 overcame these. SSH2 is open source, and it can be found at the following websites:

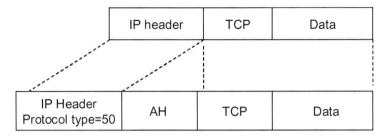

Fig. 15.10 Authentication header using transport mode

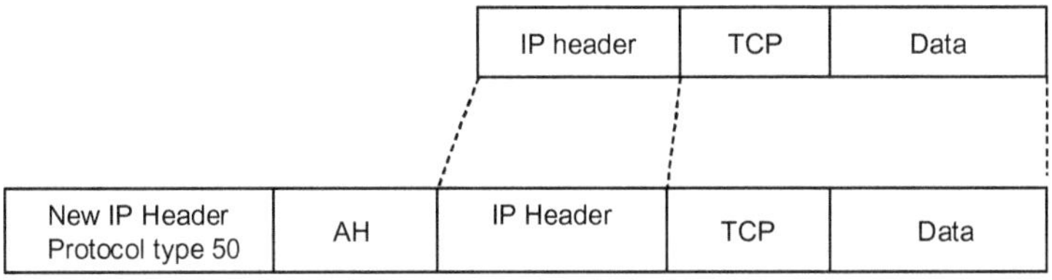

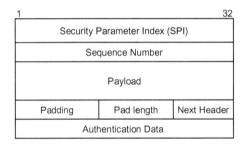

Fig. 15.11 Authentication header using tunneling mode

1			32
Security Parameter Index (SPI)			
Sequence Number			
Payload			
Padding	Pad length	Next Header	
Authentication Data			

Fig. 15.12 ESP format

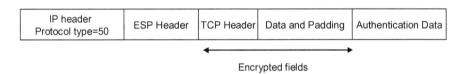

Fig. 15.13 ESP in transport mode

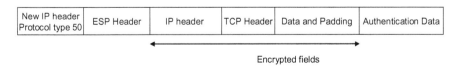

Fig. 15.14 ESP using tunneling mode

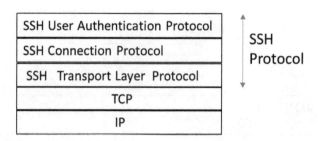

Fig. 15.15 SSH Protocol Stack

1. www.openssh.com
2. http://www.freesshd.com/
3. www.filezilla.com

Telnet is used for remote logging and FTP is used for file transfer from a remote computer. These protocols have the following weaknesses.

1. Telnet and FTP use passwords as an authentication method. When transmitted for authentication, these passwords are transmitted in plaintext.
2. Telnet and FTP transmit all data in plaintext, which does not provide protection from attacks that monitor data that can read plaintext.
3. Telnet and FTP clients do not authenticate the server.

Advantages of SSH Secure Shell   SSH offers logging into a remote device securely with the following features:

1. SSH transmits data in cipher text.
2. SSH transmits user authentication information in cipher text.
3. SSH clients authenticate the server.
4. SSH dynamically generates a key exchange between the client and the server for encryption and decryption between the server and the client.

SSH Protocol

SSH protocol is used to establish a secure connection between client and server. The objective of the creation of SSH was to replace Telnet and File Transfer Protocol. Figure 15.15 shows the SSH Protocol Stack.

SSH User Authentication Protocol (SSH-AUTH)   This verifies the client's identity by using the provided public key or password.

SSH Connection Protocol (SSH-CONNECT)   The SSH connections provide multiple channels, and these channels are multiplexed to one encrypted tunnel.

SSH Transport Protocol (SSH-TRANS)   SSH-TRANS runs on top of TCP with the port number 22. It makes a secure connection between a client and a server by providing Integrity and Confidentiality

SSH-TRANS performs the following operations:

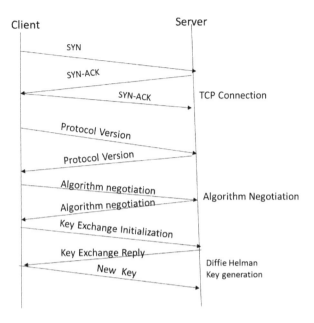

Fig. 15.16 SSH Client and Server connection establishment

1. The server and the client negotiate the SSH version and software to be used.
2. Another negotiation takes place to determine the algorithm for encryption and the MAC algorithm for integrity. Some of the used encryption algorithms are AES128-CBC, AES256-CBC, AES-CTR, and Blowfish-CBC. Some of the algorithms for Message Authentic Code (MAC) are HMAC-MD5, HMAC-SHA-256, and HMAC-SHA-512.
3. Key exchanges such Diffie–Hellman. The client and the server share a secret key and use the secret key to generate encryption and authentication keys.

Figure 15.16 shows a client and a server exchanging packets for establishing a secure connection.

15.5   IEEE 802.1X

Security is one of the most important issues in the development of Wireless LAN. In a wired LAN, users have a direct connection to the network, whereas WLAN users do not. WLAN users must verify their identity before accessing the network. Authentication is a process used by a wireless station or a wired station to identify oneself on the network. A password is used for authentication when accessing a network's resources.

Typically, the communication channel between a user and a WLAN is not secure. An attacker can monitor the communication channel and collect user data and

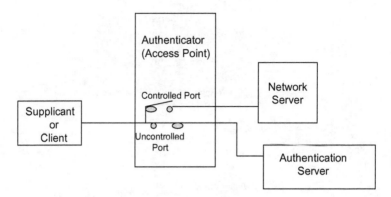

Fig. 15.17 Components of IEEE 802.1X extensible authentication protocol

passwords. The network administrator operates the access point (where a user accesses the network), but a hacker may set up a rogue access point to capture the connection of an unsuspecting client. Therefore, a method is needed such that the user can verify the authenticity of the access point. This method of authentication is called **Mutual Authentication**.

The IEEE 802.1X is a standard for port-based network access control and is an open standard for authentication of both wireless and wired stations using an authentication server. Figure 15.17 shows the components of a network employing the IEEE 802.1x standard.

Supplicant A supplicant can be any device using IEEE 802.11 protocol for networking, or any PC connected to the network.

Authenticator The authenticator can be an access point for 802.11 LAN or a switch for a wired LAN. The authenticator uses controlled and uncontrolled ports for authentication of a supplicant.

Authentication Server The authentication server performs an authentication process for a supplicant. One type of authentication server is called a **RADIUS** (Remote Authentication Dial-In User Service). A RADIUS is both an authentication server and an accounting server and is used to authorize a station on the network.

Authentication Protocol An authentication protocol is a procedure that is used by the client and the authentication server for the authentication process. IEEE 802.1x uses EAP for exchanging messages during the authentication process. EAP supports the following authentication messages between client and server:

(a) Request
(b) Response
(c) Success
(d) Failure

EAP messages between the client and the AP are carried by the EAPOL (EAP over LAN) protocol. EAP messages from the AP to the server are carried by the EAP over RADIUS protocol.

IEEE 802.1X does not define which authentication protocol to use. There are 40 authentication protocols available, and some of the most popular are as follows:

1. **EAP-MD5 (Message Digest 5)**: EAP-MD5 is a password-based authentication protocol.
2. **EAP-TLS (EAP-Transport Layer Security)**: EAP-TLS is based on mutual authentication of the client to the server and the server to the client. Both the client and the server must be assigned a digital certificate.
3. **LEAP** (Lightweight Extensible Authentication Protocol): Used by CISCO.
4. **EAP-TTLS** (Tunneled TLS).
5. **PEAP** (Protected EAP).

802.1X Operation

The authenticator (i.e., access point) contains both the logical controlled and uncontrolled ports for authentication. When using the uncontrolled logical ports, the client can communicate with the authentication server but does not have access to the network services. The following steps describe the authentication process of a client by an authentication server.

1. The client requests an association with the AP (access point).
2. The AP responds to the client's request.
3. The client sends an EAP start message to the AP.
4. The AP requests the identity of the client (such as the user name).
5. The client sends an EAP packet containing their identity to the authentication server.
6. The authentication server identifies the client.
7. The authentication server can accept or reject the client request.
8. Upon accepting the client's request, the AP sends an EAP success packet to the client and authorizes the controlled port so that the client can access the network services.

15.6 Certificates

A customer requests a web page from a stockbroker by logging into to the broker's server. The server requests an account number and a password for verification. If the information submitted by the customer is correct, then the server sends its public key to the customer for encryption. Suppose an intruder installs a system between a customer and a broker's server, as shown in Fig. 15.18. When the customer requests the broker's web page, the intruder system responds with a fake web page. The customer logs into the fake web page and the intruder sends a fake public key to the customer to use for encryption. To check the authenticity of the public key, the public key must be certified by a certificate authority.

Certificates verify the identity of a server, a program, or personal identification information such as a pictured driver license and number. A driver license is issued

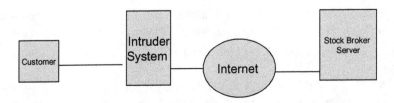

Fig. 15.18 Application of certificates

by the Department of Motor Vehicles. Digital certificates are issued in a similar fashion by the Certificate Authority (CA). Some of the certification authorities are the US Postal Service, Commercenet, and Versign.

A digital certificate contains a public key, the full name of the certificate holder, the name of the certification authority, and a digital signature of the CA. All certificates use X.509 Standard, which has been recommended by the ITU. The X.509 standard defines the information in a certificate. This information is usually divided into two sections: the data section and signature section.

The data section of X.509 contains the following information:

- **Version Number**: This number identifies the X.509 version such as V1, V2 and V3.
- **Certificate Serial Number**: This number is assigned to a certificate by the CA.
- **Signature Algorithm**: The type of algorithm that is used for the signature of a certificate.
- **Name of the Certificate Authority.**
- **Duration of the Certificate**: The period in which a certificate is valid.
- **Public Key Algorithm**: The algorithm that is used to generate a public key.

The signature section contains the following information:

- The cryptographic algorithm used by the CA to create its digital signatures.
- The digital signature of all data in the certificates.

For a user to check the validity of a certificate, he or she must perform the following functions:

1. Check the expiration date.
2. Check if the certified authority name is on the trusted list.
3. Find the digital certificates of all data and compare against those attached to the certificates.

When all of the above tests are passed, then the certificate is valid. Figure 15.19 shows a certificate.

Fig. 15.19 Image of a certificate

15.7 Firewalls

A firewall is a system that is used for preventing unauthorized users from accessing private networks. Firewalls are located between private networks and the Internet (un-trusted Network) as shown in Fig. 15.20. They can be implemented by a combination of software and hardware. Firewalls examine all packets leaving and entering a private network, blocking packets that do not meet security criteria. Firewall technologies are classified into the following types:

(a) Packet Filtering Firewall
(b) Application Proxy Server or Network Address Translation Firewall
(c) Stateful Firewall

Packet Filtering A firewall can examine each incoming packet's header such as the IP header, TCP, or UDP header, and, based on security criteria set by the network administration, accept or reject any packet. These criteria are as follows:

A. *Source IP Address and Destination IP Address*: A firewall can be programmed to block packets based on the IP address of the packet. This task is done at the Network Level of the TCP/IP model.
B. *Protocol Type*: The firewall can block packets based on protocols such as TCP, UDP, and ICMP protocols. For example, protocol filtering can filter any packet

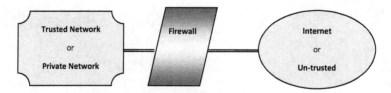

Fig. 15.20 Location of a firewall

intended for ICMP. Protocol filtering is done at the Network Level of the TCP/IP model.

C. *Source Port and Destination Port Filtering*: The firewall can block packets based on the port number of the incoming packets. The port number is used to define application protocols such as HTTP, SMTP, Telnet, and TFTP. For example, a firewall may block any incoming mail by blocking SMTP or port 25. Port filtering is done at the transport level of the TCP/IP model.

D. *Packet Payload* (*information on the payload*): A firewall can block packets based on the information contained in a packet's payloads such as a packet containing a "dirty word" or a specific sentence. The information filtering is done at the application level of TCP/IP model.

Application Proxy Server or Network Address Translation (NAT) The proxy server application is located between the trusted network and the un-trusted network. Assume a client of the private network wants to access the un-trusted network such as Internet. First, it sends a request to the proxy server and the proxy server uses its IP address to send the request, on the behalf of client, to the destination server. Therefore, destination server sees only the proxy server's IP address. The IP addresses of the clients in the private network are not exposed to the Internet. A proxy server should contain a software module for each application protocol such as HTTP, Telnet SMTP, and TFTP.

Stateful Firewall A Stateful Firewall keeps track of its connections and determines if the incoming packets belong to the current connections or not. This type of firewall can be configured such that it will deny any connection from the un-trusted network to the trusted server. The un-trusted network can only send packets to the trusted server if the trusted server requests them; otherwise, the un-requested packets will be discarded by the firewall.

Dual Firewall In any organization, there are some servers that clients from both inside and outside the network must be able to access including Web servers, DNS servers, and E-mail servers. Within these same organizations are servers which only clients from inside the organization are allowed access to, such as database servers or file servers. To protect both types of servers, two firewalls will be used as shown in Fig. 15.21. The servers that both inside and outside clients must have access to are located between two firewalls in an area known as the Demilitarized Zone (DMZ).

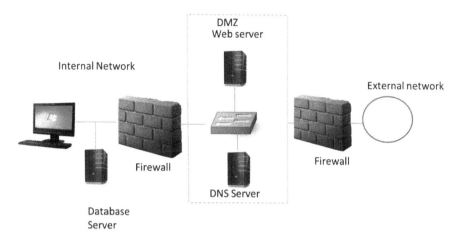

Fig. 15.21   DMZ Architecture

15.8   WLAN Security

A WLAN is less secure than a wired LAN. In a wired LAN, any person who wants access to the network must make a physical connection. In WLANs, users can access an AP if the AP signal is detectable. The following methods are used for the security of WLANs:

(a) Service Set Identifier
(b) MAC Address Filtering
(c) Wired Equivalent Privacy (WEP)
(d) Wi-Fi Protected Access (WPA), WPA2 and WPA3
(e) Authentication

Packet Sniffer
A packet sniffer is a piece of software that captures packets coming into and going out of a network. It is used for network monitoring or analysis. A packet sniffer captures all the data that passes through a network and is set to capture packets for a specific machine.

15.8.1   *Wired Equivalent Privacy (WEP)*

The IEEE 802.11 standard includes WEP in the MAC layer of WLAN to protect wireless communications. WEP uses a secret key that is shared between the AP and the users. At the transmitter side, a 24-bit Initialization Vector (IV) is appended to the secret key at the transmitter side and secret key can be 40 or 108 bits. WEP uses RC4 algorithm to generate a keystream for encryption as shown in Fig. 15.22.

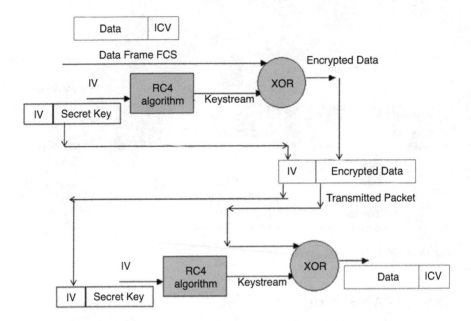

Fig. 15.22 WEP Process

The IEEE 802.11 standard defines two types of authentications: **Open System Authentication** and **Shared Key Authentication**.

Opens System Authentication
This is the simplest method of authentication for a WLAN as the access point and the client set their own criteria for authentication. If a client has the SSID of the access point, the client sends an authentication frame with its secret key to the access point. If the secret key of the client matches the secret key of the access point, then the AP sends a positive response to the client and the client becomes associated with the AP.

Shared Key Authentication Figure 15.23 shows the shared key authentication process. In shared key authentication, the clients and AP use the same key, which makes it easy for a hacker to obtain the shared key by using special sniffer software such as AirSnort or WEPcrak. Therefore, the network administrator needs to change the shared secret key frequently, which is difficult in large networks. Some of the WEP weakness are as follows: WEP does not support key management and all clients use the same shared key, WEP uses a small key size, WEP uses a 24-bit Initialization vector which is appended to a 40-bit secret key, resulting in a 64-bit WEP key and WEP does not support replay prevention.

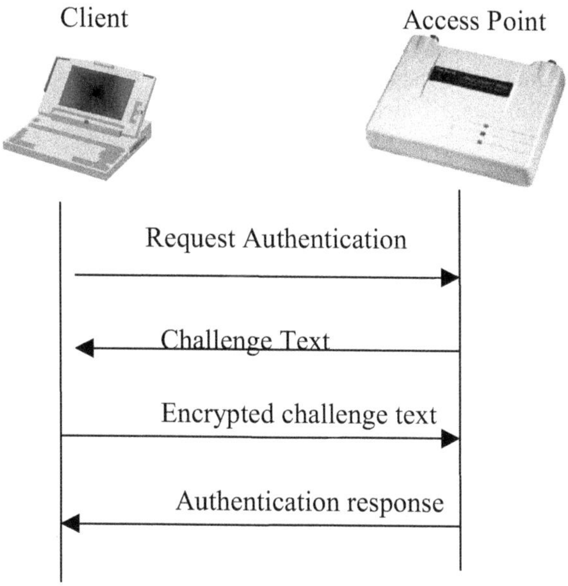

Fig. 15.23 Shared
authentication process

Client   Access Point

Request Authentication

Challenge Text

Encrypted challenge text

Authentication response

15.9   IEEE 802.11i

The IEEE802.11i-2004 amendment consisted of two sets of security protocols which are as follows:

(a) A set of protocols to overcome the weakness of WEP. It was certified by the Wi-Fi Alliance and called Wi-Fi Protected Access (WPA).
(b) A second set of security protocols called the robust security network (RSN). RSN was also certified by the Wi-Fi Alliance and called WPA2.

15.9.1   WPA (Wi-Fi Protected Access)

WPA is a subset of IEEE 802.11i and was developed to overcome the weakness of WEP by adding TKIP (Temporal Key Integrity Protocol), IEEE802.1X, Extensible Authentication Protocol (EAP), and the RADIUS Server. TKIP enhanced WEP by adding the following features:

1. Initialization vector (IV)

   (a) IV size changed from 24 bits to 48 bits.
   (b) IV created from a number sequence to avoid replay attacks.

2. For each packet a new IV is used.
3. Key management.

4. Use of the RC4 algorithm for encryption.
5. WPA uses a Message Integrity Code known as Michael. The key management and authentication protocols used by WPA and WPA2 are the same. The only difference is that WPA uses the RC4 algorithm for encryption with Michael as a Message Integrity Code, while WPA2 uses AES–CCMP for encryption and message integrity.

15.9.2   Wi-Fi Protected Access-2 (WPA2)

WPA2 operates in two modes, WPA2 Pre-shared Key (WPA2-PSK) and WPA2 Enterprise.

WPA2 Enterprise mode uses RADIUS Server and EAP for authentication. The user provides authentication information and is authenticated by an authentication server (RADIUS). Larger networks use WPA2 Enterprise mode.

WPA2 Personal mode (WPA2-PSK) provides a simple authentication method called pre-shared key for authentication and does not require a specific authentication server.

IEE802.11i ratified the Robust Security Network (RSN) for Wireless LANs. RSN includes the following:

1. IEEE 802.1x Authentication.
2. Extensile Authentication Protocol (EAP).
3. WPA2 Enterprise uses the AES algorithm with Counter Mode (CTR) for encryption and combines CTR with Cipher Block Chaining Message Authentication Code to generate the MAC. This is called AES-CCMP for encryption and message integrity.
4. Supports multiple Authentication Protocols.

Counter Mode (CTR)
The information is divided into blocks as shown in Fig. 15.24. The counter value is encrypted by AES and XORed with each block of data resulting in encrypted text, or ciphertext.

Cipher Block Chaining Message Authentication Code (CBC-MAC)
Figure 15.25 shows the Cipher Block Chaining Message Authentication Code. The ciphertexts of Fig. 15.24 are the inputs to the CBC-MAC.

WPA and WPA2 Key management
IEEE 802.11i defines a Pairwise Key for unicast transmission and a Group Key for multicast and broadcast transmission:

Pairwise Master Key (PMK): This key can be generated by a server-based key generator and transmitted to an AP and a client during an authentication operation, or as a pre-shared key.

Temporal keys consist of the following 128-bit keys:

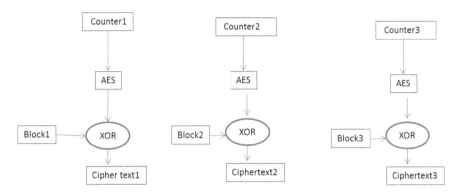

Fig. 15.24   Counter Mode Encryption

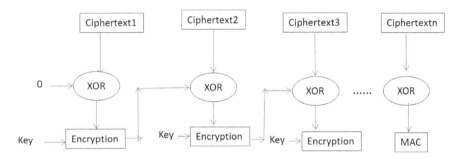

Fig. 15.25   Cipher Block Massage Authentication Code (CBC-MAC)

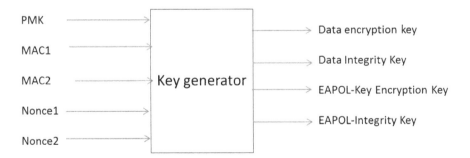

Fig. 15.26   Cipher Block Massage Authentication Code (CBC-MAC)

A. **Data Encryption Key**: This key is used for encryption.
B. **Data Integrity Key**: This key is used to generate a Message Integrity Code (MIC).
C. **EAPOL-Key Encryption Key**: This key is used for the Extensible Authentication Protocol over LAN for the authentication of a client.
D. **EAPOL-Key Integrity Key**: Used by the EAPOL protocol to check integrity.

Every time a device joins the network, the above keys are generated. Figure 15.26 shows a block diagram of the temporal key generator, which creates the four keys that make up the Pairwise Transient Key (PTK).

MAC1 and MAC2 are the MAC addresses for a client and access point.

Nonce1 and Nonce2 are random numbers which are used only once.

ASE-CCMP uses the same key for encryption and message integrity, and the PTK for WPA2 utilizes 3 keys instead of 4.

15.9.3   WPA3 (Wi-Fi Protected Access-3)

WPA3 is the latest generation of wireless network security based on IEEE802.11s. WPA3 improves the level of security used by WPA2 standard. WPA3 offers two modes of operation, WPA3 Personal (WPA-3 SAE) and WPA3-Enterprise

WPA3 Personal (WPA-3 SAE) Mode The WPA2 pre-shared key does not protect wireless networks from dictionary attacks. WPA3 Personal uses Simultaneous Authentication of Equals (SAE) for authentication process to protect against the weaknesses of WPA2 Personal from brute force/dictionary attacks. SAE offers secure password-based authentication even if the password is not complex. The SAE uses elliptic curve and Diffie–Helman to generate a Pairwise Master Key, as shown in Fig. 15.27. In the SAE commit phase, the AP and the client send their password and ECC Group (*elliptic curve cryptography*) to each other and they generate a Pairwise Master Key (PMK) then confirm the generation of PMK to each other. WPA3-Personal supports only the AES encryption mode. Table 15.1 shows a comparison of WEP, WPA, WPA2, and WPA3.

WPA3-Enterprise Mode WPA3 Enterprise is used for governments and financial institutions that require higher security. WPA3-Enterprise is similar to WPA2-Enterprise and offers optional 192-bit cryptography. The following are the list improved security features of WPA3:

Fig. 15.27 WPA3 Personnel PMK generation

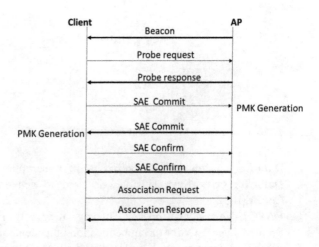

Table 15.1 Comparison of WEP, WPA, and WPA3

	WEP	WPA	WPA2	WPA3
Timeline	1997	2003	2009	2018
Authentication	Challenge Password	PSK or IEEE 802.1x	PSK or IEEE 802.1x	
Key size (bits)	40	128	128	128 bits (Personal) 192 bits (Enterprise)
Encryption	RC4 with shared key	RC4 with TKIP	CCMP with AES-128 (AES-CCMP)	AES
Data integrity	CRC-32	MIC	CBC-MAC	Secure Hash
Replay detection	No	Yes	Yes	Yes
Key management	No	Yes	Yes	Yes

- Data protection: The Suite-B 192-bit security suite is used to increase the key length.
- Key protection: The HMAC-SHA-384 algorithm is used to export keys in the four-way handshake phase.
- Traffic protection: The 256-bit Galois/Counter Mode Protocol (GCMP-256) is used to protect wireless traffic after STAs go online.
- PMF (Protect Multicast Frame): The 256-bit Galois Message Authentication Code (GMAC-256) is used to protect multicast management frames.

WPA3-Enterprise supports the following EAP cipher suites:

- TLS_ECDHE_ECDSA_WITH_AES_256_GCM_SHA384
- TLS_ECDHE_RSA_WITH_AES_256_GCM_SHA384
- TLS_DHE_RSA_WITH_AES_256_GCM_SHA384

Summary

- The Secure Socket Layer (SSL) is used for secure communication between a web browser and a web server.
- The IETF modified SSL v3 and called it Transport Layer Security (TLS).
- The SSL/TLS protocol is located between the TCP protocol layer and the application protocol.
- HTTPS uses port 443, where HTTP uses port 80 for interaction with TCP.
- The SSL/TLS protocol consists of a Handshake Layer and Record Layer.
- The function of a VPN (Virtual Private Network) device is to make secure communication between two LANs by establishing a tunnel between its two endpoints.

- VPNs provide security components such as authentication, confidentiality, and data integrity.
- IETF developed the IPsec protocol for integrity, authentication, and confidentiality of data over IP networks.
- Tunneling is the process of placing one packet inside another packet for transmission over a public network.
- IPsec is made of two protocols: Authentication Header Protocol (AHP) and Encapsulation Security Protocol (ESP).
- SSH (Secure Shell) was developed to overcome the weaknesses of Telnet and FTP.
- SSH protocol is used to establish a secure connection between client and server.
- The IEEE 802.1X is a standard for port-based network access control and is an open standard for authentication of both wireless.
- IEEE 802.1X does not define which authentication protocol to use.
- A certificate is used to identify a server to a client.
- Certificates verify the identity of a server, a program, or personal identification.
- Some of the certification authorities are the US Postal Service, Commercenet, and Versign.
- X.509 is a standard that is used in a certificate.
- A firewall is a system that is used for preventing unauthorized users from accessing private networks.
- Types of firewalls are as follows: Packet Filtering, Application Proxy Server (Network Address Translation), and Stateful Firewall.
- WLAN Security methods are follows: Service Set Identifier, MAC Address Filtering, Wired Equivalent Privacy (WEP), Wi-Fi Protected Access (WPA) WPA2, WPA3, and Authentication.

Key Terms

802.1X	Secure Shell (SSH)
Authentication Header Protocol	Secure Socket layer (SSL)
Certificates	Secure Socket Layer Protocol (SSL)
Encapsulation Security Packet (ESP)	Security Suite
Extensible Authentication Protocol (EAP)	SSH Protocol
Firewall	SSL Handshake Protocol
Firewalls	TLS/SSL
Hello Packet	Tunneling
HTTPs	Virtual Private Network (VPN)
HTTPS uses port 443	Wi-Fi Protected Access (WPA)
IEEE 802.11i	Wired Equivalent Privacy (WEP)
IEEE 802.1X Extensible Authentication Protocol (EAP)	WPA2
IP Security Protocol (IPsec)	WPA3
Mutual authentication	

Review Questions

Multiple Choice Questions

1. The Secure Socket Layer provides _____

   (a) Encryption
   (b) Authentication
   (c) Integrity
   (d) All above

2. The Secure Socket Layer is located between the _____

   (a) TCP and IP protocols
   (b) TCP and HTTP protocols
   (c) TCP and HTTPS protocols
   (d) TCP and DNS protocols

3. The SSL/TLS Client Hello contains:

   (a) Client Cipher Suite
   (b) Client Encryption Algorithm
   (c) Server Encryption Algorithm
   (d) Client Hash Algorithm

4. Which of the following is not in the SSL/TLS handshake packet?

   (a) Client Hello
   (b) Server Hello
   (c) Client Key Exchange
   (d) Server Certificate

5. VPNs are used to connect the networks _____

   (a) Located in the same building
   (b) Located in a campus
   (c) Located in different locations
   (d) Located in the same room

6. VPNs use _____ to connect networks.

   (a) Leased lines
   (b) Modems
   (c) The Internet
   (d) Public Networks

7. IEEE 802.1x is a standard for _____

   (a) Digital signatures
   (b) Authentication
   (c) Certificates
   (d) All of the above

8. Which of the following servers are usually in the DMZ?

   (a) DNS Server
   (b) Database Server
   (c) Accounting Server
   (d) Print Server

9. IPsec was developed by the _____

   (a) ITU
   (b) IETF
   (c) IEEE
   (d) EIA

10. IPsec is used for _____

   (a) Ethernets
   (b) The Internet
   (c) VPNs
   (d) VLANs

11. Which of the following is not a part of the IPsec protocol?

   (a) Authentication Header
   (b) Encapsulation Header
   (c) HTTPs
   (d) (a) & (b)

12. The X.509 is the standard for _____

   (a) HTTPs
   (b) Telnet
   (c) EAP
   (d) Certificates

13. A certificate is issued by a/the _____

   (a) Client
   (b) Server
   (c) Certificate authority
   (d) Internet

14. SSH is used to secure _____

   (a) Telnet
   (b) FTP
   (c) VLAN
   (d) (a) & (b)

15. IEEE 802.1x is used for _____

    (a) Authentication
    (b) Encryption
    (c) Secure Hash
    (d) None of the above

16. Which of the following firewalls perform NAT?

    (a) Packet filtering
    (b) Stateful
    (c) Application Proxy
    (d) None of the above

17. What methods does WPA2-Personal use for authentication?

    (a) Pre-shared Keys
    (b) IEEE 802.1X
    (c) RADIUS Server
    (d) None of the above

18. What methods does WPA2-Enterprise use for authentication?

    (a) Pre-shared Keys
    (b) IEEE 802.1X with EAP
    (c) RADIUS Server
    (d) None of the above

19. What methods does WPA2 use for encryption?

    (a) RC4
    (b) AES-CCMP
    (c) DES
    (d) ECC

Short Answer Questions

1. What does TLS stand for?
2. List the SSL/TLS protocols.
3. List the SSL handshake protocols.
4. List the information in the SSL/TLS Client Hello.
5. List information in the SSL/TLS Server Hello
6. What is an application of VPNs?
7. What is an application of IPsec?
8. List the IPsec modes of operation.
9. What does SSH stand for?
10. What are the applications of SSH?
11. List SSH protocols.
12. What is 802.1x?
13. What is EAP?

14. List three EAP protocols.
15. What is the weakness of WPA2-PSK
16. What is the name of the algorithm used for encryption by WPA2?
17. What is a nonce?
18. Show the block diagram for Counter Mode (CTR).
19. Show the block diagram of CBC-MAC.
20. What are the different types of firewalls?

Chapter 1: Introduction to Communications Networks

Selected Answers: Multiple Choice Questions

2. In a **Client/Server** network, the client submits a task to the server, then the server executes and returns the result to the requesting client station.

   (a) Peer-to-Peer
   (b) Server Based
   (c) **Client/Server**
   (d) All of the above

4. A **mail server** stores all the client's mail.

   (a) File server
   (b) Print server
   (c) Communication server
   (d) **Mail server**

6. In a **star** topology, all stations are connected to a central controller or hub.

   (a) **Star**
   (b) Ring
   (c) Bus
   (d) Mesh

8. A **hybrid** topology is a combination of different topologies connected together by a backbone cable.

   (a) Star
   (b) Ring
   (c) Bus
   (d) **Hybrid**

© The Editor(s) (if applicable) and The Author(s), under exclusive license to
Springer Nature Switzerland AG 2024
A. Elahi, A. Cushman, *Computer Networks*,
https://doi.org/10.1007/978-3-031-42018-4

10. Which type of topology uses multi-point connections? **bus**

    (a) **Bus**
    (b) Star
    (c) Ring
    (d) Full mesh

12. Which of the following networks is used for office buildings? **LAN**

    (a) **LAN**
    (b) MAN
    (c) WAN
    (d) Internet

14. The Internet is a collection of LANs connected together by **gateways.**

    (a) Routers
    (b) Switches
    (c) **Gateways**
    (d) Repeaters

16. The IEEE developed the **IEEE 802** standard for LAN.

    (a) **IEEE 802**
    (b) RS232
    (c) OSI Model
    (d) All of the above

18. The **ANSI** defines standards for programming languages.

    (e) IEEE
    (f) ISO
    (g) **ANSI**
    (h) IETF

20. Microsoft's version of IPX/SPX is called **NWLink**.

    (a) Net BEUI
    (b) TCP/IP
    (c) **NWLink**
    (d) X.25

22. The **Network** layer establishes a connection.

    (a) **Network**
    (b) Physical
    (c) Data-Link
    (d) Application

24. Which layer of the OSI model is responsible for forming a frame? **Data-Link**

   (a) **Data-Link**
   (b) Transport
   (c) Session
   (d) Physical

26. The function of the network layer is **Routing**

   (a) Error detection
   (b) **Routing**
   (c) To set up a session
   (d) Encryption

28. Which layer determines the route for packets transmitted from source to destination? **Network**

   (a) Data-Link
   (b) **Network**
   (c) Transport
   (d) Physical

Selected Answers: Short Answer

2. Explain the function of a server.
   A server is a computer on the network that holds the shared files and the network operating system that manages the network operation.

4. Explain the term "client/server model."
   A client first submits a task to the server. The server then executes the client's task and returns the results to the requesting client station.

6. What are the three components of a network?
   Network Interface Card, Transmission Medium, Network Operating System.

8. List the six networking topologies.

   • **Star**
   • **Ring**
   • **Bus**
   • **Mesh topology**
   • **Tree**
   • **Hybrid**

10. What is a hub?
   A hub links stations in the network together. The function of the hub is to accept information from one station and repeat the information to other stations or hubs.

12. What does MAN stand for?
    Metropolitan Area Network
14. What does WAN stand for?
    Wide Area Network
16. List the layers of the OSI Model.

    - **Application**
    - **Presentation**
    - **Session**
    - **Transport**
    - **Network**
    - **Data-Link**
    - **Physical**

18. List three applications of TCP/IP Model
    HTTP, FTP, Telnet
20. What layer deals with frames?
    Data-Link layer

Chapter 2: Data Communications

Selected Answers: Multiple Choice Questions

2. Modern computers work with **digital** signals.

   (a) **Digital**
   (b) Analog
   (c) (a) and (b)
   (d) None of the above

4. **Asynchronous** transmission transmits data character by character.

   (a) **Asynchronous**
   (b) Synchronous
   (c) Full duplex
   (d) Half-duplex

6. In **simplex** mode, transmission of data goes only in one direction.

   (a) **Simplex**
   (b) Half-duplex
   (c) Full duplex
   (d) Serial

8. The **bandwidth** of a communication signal is the range of frequencies that the signal occupies.

   (a) Data rate
   (b) **Bandwidth**
   (c) Baud rate
   (d) Broadband

10. Cyclic Redundancy Check can **detect one or more errors.**

    (a) Detect a single error and correct it
    (b) Detect double errors and correct them
    (c) **Detect one or more errors**
    (d) Correct more than one error

12. What is decimal value for $(111101)_2$? **61**

    (a) 44
    (b) 63
    (c) **61**
    (d) 52

14. The binary value for 45 is **101101**

    (a) 101011
    (b) **101101**
    (c) 101111
    (d) 011111

16. Asynchronous communication uses: **stop and start bits to indicate the start and end of the character**

    (a) **Stop and start bits to indicate the start and end of the character**
    (b) A start bit to synchronize transmission
    (c) Start and stop bits used for clocking
    (d) None of the above

Selected Answers: Short Answer

2. What is frequency?
Frequency is the number of cycles of a signal per second, which is measured in Hz.

4. What is the frequency of an analog signal that is repeated every .02 ms?

    $f = 1/T$
    $f = 1/.02$ ms
    $f = 1/.00002$ s
    $f = 50$ kHz

6. Sketch a digital signal.

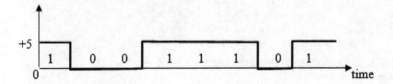

8. What is a byte?
   A byte is eight bits grouped together.
10. Convert the following binary number to Hex.
   $(111000111001)_2 = \textbf{(E39)}_{16}$
12. Convert the following number to binary.
   $(FDE6)_{16} = \textbf{(1111110111100110)}_2$
14. Convert the word NETWORK to hexadecimal.

N	E	T	W	O	R	K
4E	45	54	57	4F	52	4B

16. What is serial transmission?
   Serial transmission is a transmission method in which information is sent sequentially, bit by bit.
18. What is the advantage of parallel transmission over serial transmission.
   Parallel transmission is faster than serial transmission when both are at the same clock speed.
20. What is a synchronous transmission?
   Synchronous transmission uses an external connection, typically a clock pulse, to synchronize and set the speed of information going from a transmitter to a receiver.
22. Show the format of asynchronous transmission.

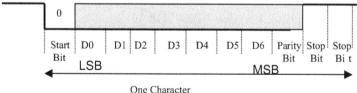

One Character

24. List two types of digital encoding methods in which clock is embedded to data signal
   Manchester Encoding and Differential Manchester Encoding.
26. List sources of error in networking.
   Crosstalk, white noise, impulse noise, and attenuation.
28. Find the BCC for word "ETHERNET."

Row even parity	ASCII Binary	Character
1	1000101	E
1	1010100	T
0	1001000	H
1	1000101	E
1	1010010	R

0	1 0 0 1 1 1 0	N
1	1 0 0 0 1 0 1	E
1	1 0 1 0 1 0 0	T
1	1 1 0 1 1 1 0	Odd parity for columns
	The BCC is 11101110	

28. Find the FCS for message 10110110 using circuit in question 29.

Input	C2	C1	C0
Initial value	0	0	0
1	0	0	1
0	0	1	0
1	1	0	1
1	0	0	0
0	0	0	0
1	0	0	1
1	0	1	1
0	1	1	0
0	1	1	1
0	1	0	1
0	0	0	1

   The FCS 001

32. **Show the digital wave form for 0101011110**

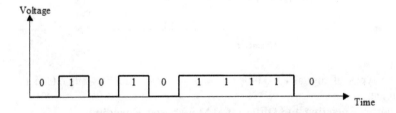

34. Calculate frequency of a signal repeated every 0.0005 seconds.

   f = 1/T
   f = 1/0.0005
   f = 2000 Hz

36. What is burst error?
   When two or more consecutive bits in a frame have changed in error.

Chapter 3: Communications Channels and Media

Selected Answers: Multiple Choice Questions

2. **Coaxial and fiber-optic** cables are used to transmit high-speed and analog signals.

    (a) UTP and coaxial
    (b) STP and UTP
    (c) **Coaxial and fiber-optic**
    (d) Fiber-optic and STP

4. Wireless transmission does not use any transmission medium.

    (a) WAN
    (b) LAN
    (c) **Wireless**
    (d) Internet

6. Which of the following UTP cables is suitable for a data rate of 100 Mbps? **Cat-5**

    (a) Cat-2
    (b) Cat-4
    (c) Cat-3
    (d) **Cat-5**

8. What type of fiber-optic cable is used for long distance transmission? **Single-mode**

    (a) Multimode graded index
    (b) **Single mode**
    (c) UTP
    (d) STP

© The Editor(s) (if applicable) and The Author(s), under exclusive license to
Springer Nature Switzerland AG 2024
A. Elahi, A. Cushman, *Computer Networks*,
https://doi.org/10.1007/978-3-031-42018-4

10. What is the maximum length that a fiber cable with a modal bandwidth of 1000 MHz*km can be in order to transmit information with a 200 MHz speed? **5 km**

    (a) 1 km
    (b) 10 km
    (c) **5 km**
    (d) 20 km

12. SONET uses byte multiplexing in **all** levels.

    (a) Upper
    (b) Mid
    (c) **All**
    (d) None of the above

14. SONET transmits the STS-1 at the rate of **8000** frames/second.

    (a) 6000
    (b) 7000
    (c) **8000**
    (d) 1000

16. STS-1 has data rate of **51.84** Mbps.

    (a) 810
    (b) 8000
    (c) **51.84**
    (d) 1.54

18. An STS-1 frame made of **9 rows and 90 columns**

    (a) 9 columns and 90 rows
    (b) **9 rows and 90 columns**
    (c) 10 rows and 100 columns
    (d) None of the above

20. An STS-3 is generated by multiplexing **three STS-1s**

    (a) **Three STS-1s**
    (b) Six STS-1s
    (c) Five STS-1s
    (d) Two STS-1s

22. An STS-3 frame format is made up of **9 rows and 270 columns**

    (a) 270 rows and 9 columns
    (b) **9 rows and 270 columns**
    (c) 10 rows and 300 columns
    (d) None of the above

Selected Answers: Short Answer

2. What does UTP stand for?
   Unshielded twisted pair.
4. Which organization defines standards for cables?
   The Electronic Industries Association (EIA).
6. What type of light source is used for fiber-optic cables?
   Light Emitting Diodes (LEDs) or Injected Laser Diodes (IJDs).
8. What are the types of fiber-optic cables?
   Single-mode fiber (SMF) and Multimode fiber (MMF).
10. What does MMF stand for?
    Multimode fiber.
12. What is the application of multimode fiber?
    Multimode fiber allows for multiple rays of light to propagate through different angles, allowing for transmission at distances less than 1 km.
14. What is the range of microwave frequencies?
    1 GHz to 23 GHz
16. What are the advantages of optical cables over coaxial cables?
    Optical cabling allows for transmission over a longer distance, greater bandwidth, increased resilience to noise, and is more difficult to tap in to.
18. Explain digital bandwidth.
    Digital bandwidth is the maximum data rate that can be transmitted over a link.
20. Find the transmission time of a packet of 1500 bytes transmitted over a 100 Mbps channel.

$$\text{Tx} = \left(1500\,\text{bytes}^* \, 8\,\text{bits}/\text{byte}\right)/100\,\text{Mbps}^* \, 10^6 \,\text{bits per second}$$
$$= 0.00012\,\text{s} = .12\,\text{ms}$$

22. 2000 bytes of data are to be transferred between a server and a host computer, which are connected via a 1000-m Cat-5 cable with a transmission rate of 10 Mbps. Calculate the following:

    (a) Transmission time

    Tx = packet size (bits) / bandwidth
    Tx = (2000 bytes 8 bits/byte) / (10 Mbps * 10^6)
    Tx = 1.6 ms

    (b) Propagation delay

    Tp = Length of communication channel (meters) / Speed of light
    Tp = 1000 m / 2 * 10^8
    Tp = .5 ms

(c) Round trip time

$$L = Tx + Tp + Tb$$
$$L = 1.6 \text{ ms} + .5 \text{ ms} + 0$$
$$L = 2.1 \text{ ms}$$
$$RTT = 2L$$
$$RTT = 2(2.1 \text{ ms})$$
$$RTT = 4.2 \text{ ms}$$

24. What is the bandwidth of a 20-km link for transmitting 500 bytes of information such that the propagation delay is equal to transmission delay?

$$Tp = Tx$$

$$\left(\text{Length of communication channel}(\text{meters})/\text{Speed of light}\right)$$
$$= \left(\text{packet size}(\text{bits})/\text{bandwidth}\right)$$

$$\left(20 \text{ km}^* 10^3 \text{ km}/\text{m}/2^* 10^8\right) = \left(500 \text{ bytes}^* 8 \text{ bits}/\text{byte}/\text{Bandwidth}\right)$$

$$.1 \text{s} = \left(4000 \text{ bits}/\text{Bandwidth}\right)$$

$$\text{Bandwidth} = \left(4000 \text{ bits}/.1 \text{s}\right)$$

$$\text{Bandwidth} = 40 \text{ Mbps}$$

26. Calculate the latency for transmitting 1500 bytes of data over the following links:

(a) 100 m copper with a bandwidth of 10 Mbps

$$L = Tp + Tx + Tb$$

$$L = \left(100 \text{ m}/2.3^* 10^8\right) + \left(\left(1500 \text{ bytes}^* 8 \text{ bits}/\text{byte}\right)/10 \text{ Mbps}^* 10^6\right) + 0$$

$$L = (.0004 \text{ ms}) + (1.2 \text{ ms}) + 0$$

$$L = 1.2004 \text{ ms}$$

(b) 4000 m optical fiber with a bandwidth of 10 Mbps

$$L = Tx + Tp + Tb$$

$$L = \left(4000 \text{ m}/2^* 10^8\right) + \left(\left(1500 \text{ bytes}^* 8 \text{ bits}/\text{byte}\right)/10 \text{ Mbps}^* 10^6\right) + 0$$

$$L = (.002 \text{ ms}) + (1.2 \text{ ms}) + 0$$

$$L = 1.202 \text{ ms}$$

28. Find the maximum data rate of a communication link with a bandwith of 3000 Hz using 8 signal levels.

$$\max \textbf{Data Rate} = 2\,W^* \log_2 N \textbf{ bps}$$

$$\max \textbf{Data Rate} = 2\left(3000\,\text{Hz}\right)^* \log_2\left(8\right)$$

$$\max \textbf{Data Rate} = 18,000\,\textbf{bps}$$

30. What does SONET stand for?
Synchronous Optical Network.
32. What is an application of SONET?
SONET is a high-speed optical data carrier.
34. What is the transmission media for SONET?
SONET uses fiber-optic cabling as transmission media.
36. List the SONET components.

- **STS Multiplexer**
- **STS Demultiplexer**
- **ADD/DROP Multiplexer**
- **Regenerator**

38. What is OC-1?
Optical Carrier Type 1.
40. How many bytes is STS-1?
810 bytes.
42. SONET transmits how many frames per second?
8000 frames per second.
44. Explain the function of Add-Drop Multiplexing.
Add/Drop multiplexing is used to extract the lower signal from the multiplexer (drop) or add a signal to the multiplexer (add).
46. Why is the STS-1 bit rate 51.84 Mbps?
The STS-1 frame is made up of 810 bytes, transmitted at the rate of 8000 frames per second:

$$810^* \, 8^* \, 8000 = 51.84\,\textbf{Mbps}$$

Chapter 4: Multiplexer and Switching Concepts

Selected Answers: Multiple Choice Questions

2. **FDM** divides the bandwidth of a transmission line into channels.

   (a) TDM
   (b) **FDM**
   (c) SPM
   (d) FSPM

4. Wave division multiplexing is used for **optical signals.**

   (a) **Optical signals**
   (b) Analog signals
   (c) Digital signals
   (d) Radio frequency signals

6. **SPM and FPM** dynamically allocate bandwidth to active inputs.

   (a) TDM and FDM
   (b) **SPM and FPM**
   (c) FPM and TDM
   (d) FDM and SPM

8. One of the applications of TDM is the **T1** link.

   (a) CABLE modem
   (b) **T1**
   (c) cable modem
   (d) LAN

© The Editor(s) (if applicable) and The Author(s), under exclusive license to
Springer Nature Switzerland AG 2024
A. Elahi, A. Cushman, *Computer Networks*,
https://doi.org/10.1007/978-3-031-42018-4

10. Pulse Code Modulation is used to convert: **analog to digital**

   (a) Digital to analog
   (b) **Analog to digital**
   (c) Digital to digital
   (d) Analog to analog

12. The bandwidth of a telephone system is **4 KHz**

   (a) 3 KHz
   (b) **4 KHz**
   (c) 8 KHz
   (d) 40 KHz

Selected Answers: Short Answer

2. List the types of multiplexers.
   - **Time Division Multiplexer**
   - **Statistical Packet Multiplexer**
   - **Frequency Division Multiplexer**
   - **Wave Division Multiplexer**
   - **Code Division Multiplexer**

4. Describe the Statistical Packet Multiplexer.
   A Statistical Packet Multiplexer dynamically allocates bandwidth to the active users.

6. What is an application of WDM?
   WDM is used for transmitting multiple rays of light simultaneously across a single fiber-optic cable.

8. What is the function of an optical transponder?
   An optical transponder is used to change the wavelength of an optical signal, such as for WDM.

10. What is the function of a Codec?
    A codec is used to convert analog signals to digital and vice versa.

12. Why must the human voice be sampled at the rate of 8000 samples per second?
    According to the Nyquist Theorem when converting analog to digital signals, an analog signal must be sampled at twice the rate of the highest frequency in the signal. The highest frequency of the human voice is generally considered to be 4000 Hz.

14. What type of Multiplexer is used in a T1 link?
    TDM.

16. What is the difference between a DS-1 and a T1 link?
    DS-1 is the frame format of the T1 link.

18. How many voice channels can be carried by a T3 Link?
   672.
20. Show the frame format of a T1 link

1 bit	Byte#24	Byte#23		Byte#2	Byte#1

22. The following inputs are connected to a 4 * 1 Statistical Multiplexer, show the outputs of the multiplexer:

a. Input #1   A - A - A
b. Input #2   B B - B B   4*1   DBA   CB   DCA   B   DBA
c. Input #3   - C C - -   MUX
d. Input #4   D- D – D

24. Use the chip sequences from Table 4.2 to find the data transmitted for Node C at the receiver side. Assume nodes A, B, and C have transmitted the following data:

Node A	111
Node B	010
Node C	001

Node A	−1 −1 −1 −1	−1 −1 −1 −1	−1 −1 −1 −1
Node B	+1 −1 +1 −1	−1 +1 −1 +1	+1 −1 +1 −1
Node C	−1 −1 +1 +1	−1 −1 +1 +1	+1 +1 −1 −1
Sum	−1 −3 +1 −1	−3 −1 −1 +1	+1 −1 −1 −3
Node C's chip bit	+1 +1 −1 −1	+1 +1 −1 −1	+1 +1 −1 −1
Node C data	0	0	1
Inner product of Node C's chip bit and Sum	(−1 −3 −1 +1) /4	(−3 −1 +1 −1)/4	(+1 −1 +1 +3) /4
Results	−1	−1	1

Chapter 5: Error and Flow Control

Selected Answers – Short Answer

2. List three Data Link layer protocols.
    SDLC. HDLC and LAPB.
4. What are the types Continuous ARQ flow control?
    Go-Back-N ARQ and Selective Reject ARQ.
6. Show the frame format of IEEE 802.2.

	1 byte	1 byte	1 or 2 bytes	
	DSAP	SSAP	Control	Information

8. What is a frame transmission method?
    A frame transmission method is the specific method that transmitters and receivers use to transmit frames. An example is bit-oriented transmission, where start and end of frame sequences are used to differentiate control information from real data.
10. What is an application of HDLC?
    HDLC is a Data Link layer link access protocol standardized by the OSI.
12. Explain the function of synchronization bits.
    Synchronization bits are used to identify the start and end of a frame.
14. Explain Go-Back-N ARQ.
    Go-Back–N ARQ has the transmitter continuously transmit while the receiver acknowledges each frame in a different channel.

© The Editor(s) (if applicable) and The Author(s), under exclusive license to
Springer Nature Switzerland AG 2024
A. Elahi, A. Cushman, *Computer Networks*,
https://doi.org/10.1007/978-3-031-42018-4

16. Show the transmitted frame after bit insertion for following frame:

011111110000000011111011111110
Bit-insertion will add the following zeroes:
011111(0)11000000011111(0)011111(0)110

Chapter 6: Modulation Methods, Cable Modems, and FTTH

Selected Answers: Multiple Choice Questions

2. Cable TV is designed to transmit a **broadband** signal.

   (a) Baseband
   (b) **Broadband**
   (c) Digital
   (d) Optical signal

4. QAM modulation is a combination of **PSK and ASK.**

   (a) ASK and FSK
   (b) ASK and PSK
   (c) PSK and FSK
   (d) **None of the above**

6. What type of modulation is used in cable modems for upstream transmission? **QPSK**

   (a) QAM
   (b) DMT
   (c) **QPSK**
   (d) FSK

8. What is the lowest frequency of TV Channel 2? **54 MHz**

   (a) 40 MHz
   (b) **54 MHz**
   (c) 60 MHz
   (d) 30 MHz

© The Editor(s) (if applicable) and The Author(s), under exclusive license to
Springer Nature Switzerland AG 2024
A. Elahi, A. Cushman, *Computer Networks*,
https://doi.org/10.1007/978-3-031-42018-4

Selected Answers: Short Answer

2. Explain the function of a modem.
   A modem converts analog signals to digital and vice versa. This is useful for linking computers over preexisting telephone or cable connections, which are analog.
4. Define baud rate.
   Baud rate is the number of signals which a modem can transmit in 1 second.
6. Explain FSK modulation.
   In FSK modulation, or Frequency Shift Key modulation, changes in frequency are used to represent the bits of the signal.
8. The speed of modem is represented as **a baud rate or data rate.**
10. Draw a constellation diagram for 32 QAM using 2 amplitudes.

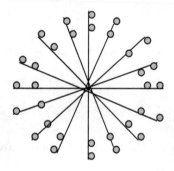

12. What does HFC stand for?
    HFC stands for Hybrid Fiber Cable, which is a combination of optical fiber and coaxial cabling.
14. What type of modulation is used in cable TV modems for upstream transmission?
    Quadrature Phase Shift Keying is used as a modulation method for upstream transmission.
16. What type of NIC is used in a computer connected to cable TV Modem?
    A 10Base-T NIC card.
18. What is data rate of a QAM signal with baud rate of 1200 and each signal represented by 4 bits?
    1200 signals per second * 4 bits per signal = 4800.
20. How many bits per signal can be represented by a 32 QAM signal?
    $32 = 2^5 = 5$ bits.
22. What does FTTH stand for?
    Fiber to the Home.
24. What is function of an optical splitter?
    An optical splitter splits accepts an optical signal and splits it into a number of other optical signals.

Chapter 7: Ethernet Technology

Selected Answers: Multiple Choice Questions

2. The IEEE standard for Ethernet is **IEEE 802.3**.

   (a) **IEEE 802.3**
   (b) IEEE 802.4
   (c) IEEE 802.5
   (d) IEEE 802.2

4. A destination address in an Ethernet frame has **6** bytes.

   (a) 2
   (b) 3
   (c) **6**
   (d) 8

6. Fast Ethernet uses the **IEEE 802.3u** standard.

   (a) IEEE 802.2
   (b) IEEE 802.5
   (c) **IEEE 802.3u**
   (d) IEEE 802.4

8. The role of the **LLC sublayer** is to interface the MAC sublayer to the physical medium dependent.

   (a) 100BaseT4
   (b) 100BaseTX
   (c) Convergence sublayer
   (d) **LLC sublayer**

© The Editor(s) (if applicable) and The Author(s), under exclusive license to
Springer Nature Switzerland AG 2024
A. Elahi, A. Cushman, *Computer Networks*,
https://doi.org/10.1007/978-3-031-42018-4

10. The data rate of 100Base-TX is **100** Mbps.

    (a) **100**
    (b) 10
    (c) 200
    (d) 1000

12. There are **two** types of repeaters.

    (a) Five
    (b) **Two**
    (c) Three
    (d) Four

14. Fast Ethernet's data rate is **100** Mbps.

    (a) **100**
    (b) 10
    (c) 400
    (d) 200

16. The data rate of Gigabit Ethernet is **1000** Mbps

    (a) 100
    (b) **1000**
    (c) 200
    (d) 10,000

18. Gigabit Ethernet uses **CSMA/CD** access method for half-duplex operation.

    (a) **CSMA/CD**
    (b) Token passing
    (c) Demand priority
    (d) None of the above

20. 1000BaseFX uses **fiber-optic** cabling for the transmission of data.

    (a) UTP
    (b) **Fiber-optic cable**
    (c) coaxial
    (d) STP

22. Gigabit Ethernet can operate in **full duplex and half duplex.**

    (a) Full duplex
    (b) Half duplex
    (c) **(a) and (b)**
    (d) None of the above

24. What is the transmission medium for 1000BaseT? **Cat-5 UTP**

   (a) **Cat-5 UTP**
   (b) Cat 4 UTP
   (c) Coaxial cable
   (d) Fiber cable

26. What type of fiber cable is used for Gigabit Ethernet? **Multimode and single-mode fiber.**

   (a) Multi-Mode fiber
   (b) Single-Mode
   (c) **Both a and b**
   (d) None of the above

28. The data rate of 10 GbE is **10,000** Mbps

   (a) 100
   (b) 1000
   (c) **10,000**
   (d) 100,000

Selected Answers: Short Answer

2. What do UTP and STP stand for?
   Unshielded Twisted Pair and Shielded Twisted Pair.
4. Show the Ethernet II frame format and the function of each field.

Starting Delimiter (1 byte)	Destination Address (6 bytes)	Source Address (6 bytes)	Protocol Type (2 bytes)	Information field 46–1500	Frame Check Sequence (4 bytes)

Starting Delimiter – indicates the start of a frame.
Destination Address – the destination MAC address.
Source Address – the source MAC address
Protocol Type – the protocol in use
Information – generic data being sent.
Frame Check Sequence – error detection.

6. Show the IEEE 802.3 frame format and function of each field.

	1 byte		1 byte	1-2 bytes	42-1497 bytes
	DSAP		SSAP	Control	Information

LLC Frame Format

Bytes	1	6	6	2		0-1500	0-46	4
	SFD	DA	SA	Length		Data Unit	Pad	FCS

Start of Frame Delimiter – indicates the start of a frame.
Destination Address – the destination MAC address.
Source Address – the source MAC address.
Length – the length of the entire frame.
Destination Service Access Point – used by the LLC to determine the desti-nation protocol.
Source Service Access Point – used by the LLC to determine the source protocol.
Control – determines the type of data in the information field.
Infromation – generic data being sent.
Pad – used to pad out the frame if the information field is less than 46 bytes.
Frame Check Sequence – error detection.

8. What does CSMA/CD stand for?
   Carrier Sense Multiple Access with Collision Detection.
10. What is a MAC Address?
    A MAC address is a unique 48-bit address assigned to each network inter-face card.
12. What is a jam signal?
    A jam signal is sent when a collision is detected when two stations transmit at the same time. The CSMA/CD jam signal is 32 bits of all ones.
14. Describe unicast addresses.
    A unicast address is a type of address that indicates that the recipient is a single station.
16. What is the application of CRC (Cyclic Redundancy Check)?
    CRC is used to check for errors and verify the integrity of a packet.
18. How many bits of a network address represent the manufacturer ID?
    Of the 48-bit MAC address, 22 bits are used to represent the manufac-turer ID.
20. What happens when two or more computers simultaneously transmit frames on an Ethernet network?
    This action will cause a collision, and according to CSMA/CD, one of the stations will then send a jam signal to warn all other stations as such.
22. What is the function of the back-off algorithm in an Ethernet network?
    The back-off algorithm is used with CSMA/CD after a jam is detected, and it generates waiting times that the stations use to allow them to begin trans-mission again.

24. What is function of the protocol type field in the Ethernet II frame format.
    The protocol type field indicates what protocol is in use, such as ARP.
26. What is function of the pad field in the IEEE 802.3 frame format?
    The pad field is used to pad extra bits if the total Ethernet's data field is lower than 46 bytes.
28. Explain the following terms:

    (a) 100Base4T
        100 Mbps, baseband, 4 pair Cat-3 cabling
    (b) 100BaseTX
        100 Mbps, baseband, Cat-5 cabling
    (c) 100BaseFX
        100 Mbps, baseband, fiber-optic cabling
30. What is the difference between 100BaseTX and 100BaseT4?
    100BaseTX uses Cat-5 cabling, while 100BaseT4 uses Cat-3 cabling.
32. What is the application of a Class II repeater?
    A class II repeater takes an input signal and repeats it out of its multiple ports.
34. Name the IEEE committee that developed the standard for Fast Ethernet.
    IEEE 802.3u.
36. What are the types of media used for Fast Ethernet?
    Fast Ethernet uses UTP cabling and fiber-optic cabling.
38. What type of signal encoding is used for 100BaseFX?
    NRZ-I
40. What is the IEEE standard number for Gigabit Ethernet?
    IEEE 802.3z.
42. What type of frame is used by Gigabit Ethernet?
    Gigabit Ethernet uses the IEEE 802.3 frame format.
44. List the kinds of transmission media used for Gigabit Ethernet.
    UTP cabling and fiber-optic cabling.
46. What are the hardware components of Gigabit Ethernet?
    Gigabit Ethernet connections use UTP or fiber-optic cabling to connect to gigabit switches in order to transmit data to other stations with gigabit NICs.
48. Explain the following terms:

    (a) 10GBASE-SR

        10 gigabit, baseband, short wavelength, LAN connection

    (b) 10GBASE-SW

        10 gigabit, baseband, short wavelength, WAN connection

    (c) 10GBASE-LR

        10 gigabit, baseband, long wavelength, LAN connection

(d) 10GBASE-LW

   10 gigabit, baseband, long wavelength, WAN connection

(e) 10GBASE-ER

   10 gigabit, baseband, extended long wavelength, LAN connection

(f) 10GBASE-EW

   10 gigabit, baseband, extended long wavelength, WAN connection

Chapter 8: LAN Interconnection Devices

Selected Answers: Multiple Choice Questions

2. **Switches** operate in the data link layer.

   (a) Hubs
   (b) Repeaters
   (c) **Switches**
   (d) Gateways

4. In a **source routing bridge**, the frame contains the entire route to the destination.

   (a) **Source routing bridge**
   (b) Learning bridge
   (c) Repeater
   (d) Gateway

6. A **gateway** operates up to the application layer.

   (a) Router
   (b) Switch
   (c) **Gateway**
   (d) Repeater

8. A **gateway** is used to convert one protocol to another protocol.

   (a) Router
   (b) Switch
   (c) **Gateway**
   (d) Hub

10. A switch is a device with **multiple** port(s).

   (a) Single
   (b) Two
   (c) **Multiple**
   (d) None of the above

12. Layer 3 switches or routing switches work on the OSI physical layer, data link layer and **network** layer.

   (a) Application
   (b) Session
   (c) Presentation
   (d) **Network**

14. What type of switch is used to connect LAN segments within the same network? **Layer 2 switch**

   (a) **Layer 2 switch**
   (b) Layer 3 switch
   (c) Layer 4 switch
   (d) All of the above

16. What type of switch is also used to route packets? **Layer 3 switch**

   (a) Layer 2 switch
   (b) **Layer 3 switch**
   (c) Layer 4 switch
   (d) None of the above

Selected Answers: Short Answer

2. What is the function of a repeater?
   The repeater accepts traffic from its input and then retransmits the traffic at its output, usually in order to expand the network's diameter. A hub is a multiple output repeater.

4. What layer of the OSI model does a bridge operate at?
   Bridges operate at the Data Link layer.

6. Explain the operation of a source routing bridge.
   A source routing bridge routes a frame based on the information in the routing field of the frame.

8. Explain a static router.
   A static router requires the routing table to be manually configured by a network administrator.

10. A router works in which layer of the OSI model?
   A router operates at the Network layer.

12. What is the application of a gateway?

    A gateway connects networks communicating with different protocols, such as by connecting an IEEE 802.3 network to one using DECNet.

14. What is the difference between a gateway and a router?

    A router works at the Network layer and is used to route packets to other networks. A gateway works at the Application layer, and is used to connect networks operating with different protocols.

16. What is the application of a symmetric switch?

    Symmetric switches connect LAN segments with similar data rates.

18. What does VLAN stand for?

    Virtual Local Area Network.

20. Suppose a company has two working groups, A and B. Group A has 4 computers and group B has 3 computers; all connected to an eight port Ethernet switch. Both groups need to access a common file server FS1. There is an in-house requirement that group A computers should not be able to see Group B computers in the Network.

    (a) **Draw a diagram showing an Ethernet switch with seven computers and file server.**

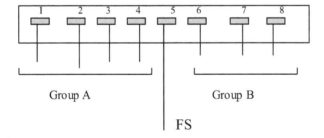

(b) Show the VLAN connectivity matrix for the above requirements.

Port #	1	2	3	4	5	6	7	8
1	+	+	+	+	+	−	−	−
2	+	+	+	+	+	−	−	−
3	+	+	+	+	+	−	−	−
4	+	+	+	+	+	−	−	−
5	+	+	+	+	+	+	+	+
6	−	−	−	−	+	+	+	+
7	−	−	−	−	+	+	+	+
8	−	−	−	−	+	+	+	+

Chapter 9: Internet Protocols Part I

Selected Answers: Multiple Choice Questions

2. The Internet uses the **TCP/IP** protocol.

    (a) X.25
    (b) NWLink
    (c) **TCP/IP**
    (d) Window NT

4. **IP** provides packet delivery between networks.

    (a) **IP**
    (b) TCP
    (c) X.25
    (d) ARP

6. **TCP** performs reliable delivery of data.

    (a) IP
    (b) UDP
    (c) **TCP**
    (d) RARP

8. What type of switching is used in the Internet? **Packet switching.**

    (a) Virtual circuit
    (b) **Packet switching**
    (c) Circuit switching
    (d) Message switching

© The Editor(s) (if applicable) and The Author(s), under exclusive license to
Springer Nature Switzerland AG 2024
A. Elahi, A. Cushman, *Computer Networks*,
https://doi.org/10.1007/978-3-031-42018-4

10. Telnet uses which of the following protocols for remote login: **TCP**

   (a) UDP
   (b) **TCP**
   (c) IP
   (d) FTP

12. Telnet enables a user to **remote login.**

   (a) Transfer a file
   (b) Send E-mail
   (c) **Remote login**
   (d) Transfer mail

14. What is the application of a loopback address? **Used for testing.**

   (a) Reserved by Internet authority
   (b) **Used for testing**
   (c) Used for broadcast address
   (d) Used for unicast

16. What protocol is used for the World Wide Web? **HTTP**

   (a) TCP/IP
   (b) **HTTP**
   (c) UDP
   (d) ARP

18. What is the function of the source and destination port in a TCP header?

   (a) It is used to identify the source and destination host on the network.
   (b) **It is used to identify the application source protocol and application of destination protocol.**
   (c) It is used to identify source protocol and destination protocol.
   (d) None of the above.

20. What is the function of Time-to-Live (TTL) in a TCP header? **It defines the number of routers a datagram can pass.**

   (a) It holds time of the day.
   (b) **It defines the number of routers a datagram can pass.**
   (c) It defines transmission time of a datagram between the source and destination.
   (d) It defines the number of words in a packet.

22. How many bits is IPv6? **128 bits**

   (a) 32 bits
   (b) 48 bits
   (c) 64 bits
   (d) **128 bits**

24. What is the function of an IP sub-net mask?

   (a) **IP sub-net mask represents the bits in the network portion of IP address.**
   (b) IP sub-net mask represents the bits in host portion of IP address.
   (c) a and b
   (d) None of the above

Selected Answers: Short Answer

2. What is the protocol used by the WWW to transfer a file?
   Hypertext Transfer Protocol (HTTP).
4. List the DNS indicators.
   com, edu, net, org, int, mil, and gov.
6. List the protocols in the Internet level.
   IP, TCMP, ARP, and RARP.
8. What is the size of an IP header?
   20 bytes.
10. Explain the function of TCP.
    The TCP performs reliable delivery of packets by adding a sequence number and ACK number to the packets transmitted.
12. What is the function of the TTL field in an IP header?
    This field specifies the number of routers that the datagram can pass through to get to the destination before being dropped.
14. List the IP address classes.
    Class A, B, C, D, and E.
16. What is the name of the organization applied to for IP addresses?
    Internet Network Information Center (InterNIC).
18. What is the newest version of IP?
    IPv6.
20. Convert the following IP address from hexadecimal to dotted decimal representation and find the class type of each IP address.

    (a) 46EF3A94
       70.239.58.148 – class A
    (b) 23446FEC
       35.68.111.236 – class A
22. The following E-mail address is given, identify the username and mail server address:
    Elahi@Xycorp.com
    The username is Elahi, the mail server address is Xycorp.
24. List three application protocols for TCP.
    FTP, SMPT, and HTTP

26. What is the function of ICMP?
    Internet Control Message Protocol is used for handling error messages and control messages.
28. Explain the function of telnet.
    Telnet is one of the application protocols of TCP/IP and is used for remote connection of one computer to another computer.
30. What is an MTU?
    Maximum Transfer Unit (MTU) is the largest possible frame size that can be sent through the Internet.
32. How many bits is IPv6?
    IPv6 addresses are 128 bit.
34. There are two computers, A and B, with IP addresses of 174.20.45.37 and 174.20.67.45. If these two computers have a sub-net mask ID of 255.255.0.0, can you determine if these two computers are in the same network?
    Yes. The subnet mask shows that these are both class B addresses, belonging to the 174.20.x.x network.

Chapter 10: Internet Protocols Part II and MPLS

Selected Answers: Multiple Choice Questions

2. How many root DNS servers exist? **13.**

   (a) 1
   (b) 10
   (c) 12
   (d) **13**

4. Which of the following protocol is used for DNS? **UDP and TCP.**

   (a) UDP
   (b) TCP
   (c) ARP
   (d) **(a) & (b)**

6. A **DHCP server** holds a range of IP addresses.

   (a) Client server
   (b) **DHCP server**
   (c) Router
   (d) Relay agent

8. DHCP Discovery packets are transmitted as **broadcast** packets

   (a) **Broadcast**
   (b) Unicast
   (c) Multicast
   (d) (a) & (b)

© The Editor(s) (if applicable) and The Author(s), under exclusive license to
Springer Nature Switzerland AG 2024
A. Elahi, A. Cushman, *Computer Networks*,
https://doi.org/10.1007/978-3-031-42018-4

10. The function of a router is to **find the path to a destination for transporting packets.**

    (a) Find the destination IP addresses
    (b) Find the destination MAC addresses
    (c) **Find the path to a destination for transporting packets**
    (d) None of the above

12. Dynamic routing tables are updated by **automatic methods.**

    (a) A server
    (b) The network administrator
    (c) **Automatic methods**
    (d) None of the above

14. A socket is a combination of a port number and **IP address.**

    (a) **IP address**
    (b) Hostname
    (c) Default gateway
    (d) (a) & (b)

Selected Answers: Short Answer Questions

 2. List the types of DNS queries.
    DNS queries can be iterative or recursive.
 4. What is the transport layer protocol used for DNS?
    DNS uses UDP as the transport layer protocol.
 6. What commands are used to display the DNS cache table?
    arp-a will show the DNS cache table.
 8. List at least 5 TLDs.
    .com, .edu, .net, .gov, .org
10. Does an organization have separate DNS servers for IPv4 and IPv6?
    No, a single DNS server can be configured for both IPv4 and IPv6.
12. List the components of DHCP.
    DHCP requires a DHCP server, DHCP client, and potentially a DHCP relay agent.
14. List DHCP packets and explain the function of each.
    DHCP Discover – a broadcast sent by a client looking for a DHCP server.
    DHCP Offer – the DHCP server sees the discover packet and responds to the request with an offer for an IP address.
    DHCP Request – the client gets the offered IP address and accepts the address and leasing term.
    DHCP Acknowledge – the server acknowledges that the client has accepted the offer, and officially leases the IP address to that client.

16. List the link characteristics.
    Hop count, throughput, communications cost, and delay.
18. What is the function of Internet Control Message Protocol (ICMP)
    ICMP is used to transport control messages, such as pings and error messages, across the IP protocol for diagnostics and troubleshooting.
20. What is the application of a socket?
    A socket is a binding of an IP address and a port number used to send and receive data.
22. Socket is combination of IP address and **port number.**

Chapter 11: Voice over Internet Protocols (Voice over IP)

Selected Answers: Multiple Choice Questions

2. One of the factors that plays an important role in successful VoIP is:

   (a) Cost
   (b) **Quality of service**
   (c) Speed
   (d) Delay

4. The Internet Engineering Task Force approved _____ for VoIP.

   (a) TCP
   (b) **SIP**
   (c) H.323
   (d) RTP

6. Real time protocol is used for_____.

   (a) Transporting data
   (b) Transporting voice
   (c) **Transporting audio and video packet**
   (d) Transporting images

8. A _____ accepts a SIP user agent request and forwards it to another user agent.

   (a) SIP endpoint
   (b) SIP gateway
   (c) **SIP proxy server**
   (d) SIP Redirector server

© The Editor(s) (if applicable) and The Author(s), under exclusive license to
Springer Nature Switzerland AG 2024
A. Elahi, A. Cushman, *Computer Networks*,
https://doi.org/10.1007/978-3-031-42018-4

Short Answer Questions

2. Does VoIP reduce or increase costs of voice communications?
   VoIP reduces the cost of voice communications.
4. What is the most important factor to consider in VoIP?
   Quality of service.
6. Define transmission delay and describe what causes it.
   Transmission delay is the time that it takes a packet to reach its destination. This can be affected by the link's data rate, the distance to the destination, and buffering time.
8. Define packet loss and describe what causes it.
   Voice packets are transmitted over UDP, and therefore there is no guarantee that packets will reach their destinations which may contribute to loss. Packets may also be dropped by gateways when there is congestion in the network.
10. What does PCM stand for and what is the function of PCM?
    PCM stands for Pulse Code Modulation, which is used to convert voice signals to digital signals.
12. What are the two protocols used for VoIP?
    SIP and H.323 are used for VoIP.
14. What is a voice compression device called?
    A voice compression device is called a codec.
16. What is the difference between a stateful and a stateless SIP proxy server?
    The stateless proxy server does not keep any information about the forwarded responses or requests. The stateful proxy server keeps information about responses and requests.
18. Describe the SIP connection operation.
    Refer to Fig. 11.5
20. Find the minimum bandwidth requirement of a VoIP channel using a 20 ms audio frame and a G.729 codec.
    G.729 compressed voice packets have a data rate of 8 Kbps, and with 20 ms of audio there is a payload of 8000 * 20/1000 = 160 bits or 20 bytes.

    The full voice packet consists of a 20-byte payload + 40-byte (RTP, UDP, and IP) header + 26-byte Ethernet header, which makes for a total of 86 bytes. The voice packet must reach 100 packets per second, meaning the bandwidth must be at least:

$$86^* 8^* 100 = 68.8 \, Kbps$$

Chapter 12: Wireless Local Area Network (WLAN)

Selected Answers: Multiple Choice Questions

2. An advantage of using spread spectrum signals over narrowband signals is that: **spread spectrum uses a range of frequencies.**

   (a) Spread spectrum has more power.
   (b) **Spread spectrum signals use a range of frequencies.**
   (c) Narrowband signals use a range of frequencies.
   (d) Spread spectrum uses a single frequency.

4. Which of the following technologies are used for WLAN? **Infrared and radio frequencies.**

   (a) Infrared
   (b) Radio frequency
   (c) **(a) & (b)**
   (d) Digital signal

6. IEEE 802.11g offers data rates of **54 Mbps.**

   (a) 2 Mbps
   (b) 1, 2, and 11 Mbps
   (c) **54 Mbps**
   (d) 11 and 45 Mbps

8. IEEE 802.11g operates in the **ISM** band.

   (a) U-NII
   (b) **ISM**
   (c) B
   (d) C

10. IEEE 802.11g uses **DSSS** and **OFDM** for transmitting information.

    (a) DSSS and FHS
    (b) **DSSS and OFDM**
    (c) DSSS and CCK
    (d) FHS and OFDM

12. DSSS uses **11** chip bits.

    (a) **11**
    (b) 12
    (c) 15
    (d) 20

Selected Answers: Short Answer

2. What are the components of a WLAN?
   A WLAN consists of a WLAN NIC, an Access Point, and a Network Operating System.
4. What is a cell?
   A cell is the area covered by an Access Point.
6. Explain narrowband signals and spread spectrum signals.
   Narrowband signals only cover a narrow range of frequencies, while spread spectrum signals cover a large range of frequencies.
8. What are the advantages of a spread spectrum signal over a narrowband signal?
   A spread spectrum signal is more difficult to jam and intercept, and noise is less disruptive in spread spectrum signals.
10. What does OFDM stand for?
    Orthogonal Frequency Division Multiplexing.
12. What does DSSS stand for?
    Direct Sequence Spread Spectrum.
14. What are the data rates for IEEE 802.11b?
    5.5 and 11 Mbps.
16. What is the maximum data rate for IEEE 802.11a?
    54 Mbps.
18. What are the types of access methods for WLANs?
    The access methods are Carrier Sense Multiple Access with Collision Avoidance and Point Coordination Function.
20. What is the function of the association request frame?
    An association request frame is used by a client to request access to a BSS network.
22. What is the Service Set identifier?
    A service set identifier

24. What is the function of Wi-Fi?
    Wi-Fi is used for wireless connection to a WLAN.
26. Explain multi-path fading.
    Multi-path fading happens when transmitted media takes multiple paths to reach the destination, ultimately reaching the destination with different delays and amplitudes.
28. Explain access methods for WLAN.
    Carrier Sense Multiple Access with Collision Avoidance will listen to the medium to become clear to avoid collisions. Point Coordination Function will have the transmitter send a beacon frame and poll clients directly for transmissions.
30. What is the maximum number of transmitters and receivers for IEEE 802.11n?
    For IEEE 802.11n, there are a maximum of 4 transmitters and 4 receivers.
32. What are the channel frequencies in which IEEE 802.11n can operate?
    IEEE 802.11n can operate in the 2.4 and 5 Ghz bands.
34. What is the advantage of frame aggregation?
    Frame aggregation results in less overhead, as multiple frames are combined under the same header.
36. What type of frame aggregation is used for single destination and multiple applications?
    A-MPDU is used for a single destination and different applications.
38. What is the guard interval?
    The guard interval is a time required between the transmission of different frames.
40. What is the Maximum data rate of IEEE802.11n?
    600 Mbps.

Chapter 13: Low Power Wireless Technologies for Internet of Things (IoT)

Selected Answers: Short Answer

2. List the name of the Low Power Wide Area Network Technology that covers more than 100 m.
   ZigBee.

4. What is the maximum number of nodes that can be used in a ZigBee Network?
   ZigBee can support up to 65,000 nodes in a single network.

6. List the ZigBee topologies.
   ZigBee uses star, tree, and mesh topologies.

8. List ZigBee device types and explain the function of each device.
   Coordinator – the coordinator starts and oversees the network, allocating addresses, and permitting devices to join and leave.
   Router – this device expands network coverage, sharing responsibilities with the coordinator except for being able to start a new network.
   Trust Center – the trust center provides authentication and security key distribution.
   End Node – the end nodes are the end devices that make up the clients in the ZigBee network.

10. What is IEEE 802.15.4?
    IEEE 802.15.4 is the standard for the MAC and physical layers of a low power personal area network, such as is in use with ZigBee.

12. List the functions of the ZigBee physical layer.
    The ZigBee physical layer receives frames and converts them to RF signals, as well as takes in RF signals for conversion to frames.

14. Show the ZigBee physical layer frame format.

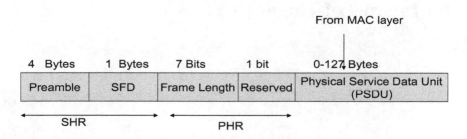

16. How many bits are addressed to a ZigBee end device?
    A ZigBee end device gets a 16-bit address.
18. What is the difference between a full function device and a reduced function device?
    A full function device operates in the full IEEE 802.15.4 layer, while a reduced function device operates with a more limited functionality.
20. What method does ZigBee use for encryption?
    ZigBee uses Counter Mode with 128-bit AES for the encryption of general network messages.
22. List the ZigBee security modes.
    ZigBee has a standard mode for security, which uses network keys provided by the trust center for encryption and decryption, but it also has a high security mode, in which entities must authenticate with each other and specific permissions tables can be established.
24. What is the application of 6LoWPAN?
    6LoWPAN is used to support IPv6 packets for Low Power Wireless Devices in support of the Internet of Things.
26. What is the function of the Adaptation Layer in 6LoWPAN?
    The Adaptation Layer is used to support compression and fragmentation of IPv6 packets.
28. List four applications for LoRa WAN.
    LoRa WAN can be applied to improve irrigation control, street lighting, wearable medical devices, and electric meters.
30. What frequency band does LoRa operate?
    LoRa operates in the ISM band.
32. What is the data rate of LoRa WAN?
    LoRa WAN has a data rate of 5 kbps in the 125 kHz band and 10 kbps in the 250 kHz band.

34. Show the LoRa WAN Protocol Architecture.

Application			
LoRa MAC			
MAC Options			
Class A	Class B	Class C (Continuous)	
LoRa Modulation			
Regional ISM band			
EU 868 MHz	EU 433 MHz	US 915 MHz	AS 430 MHz

MAC Layer

Physical Layer

Chapter 14: Introduction to Cryptography

Selected Answers: Multiple Choice Questions

2. Plaintext encrypted is called a/an **ciphertext.**

   (a) Cipher
   (b) Encryption
   (c) **Ciphertext**
   (d) Hash value

4. The science of encryption and decryption is called **cryptography.**

   (a) Authentication
   (b) Cipher
   (c) **Cryptography**
   (d) Hash value

6. Stream ciphers encrypt information **one bit at a time.**

   (a) **One bit at a time**
   (b) One byte at a time
   (c) One block at time
   (d) All at the same time

8. Computer A uses Computer B's public key for encrypting its information and sends it to Computer B. What does Computer B use for decryption? **Computer B uses its private key for decryption.**

   (a) Computer B uses its public key for decryption.
   (b) **Computer B uses its private key for decryption.**
   (c) Computer B uses Computer A's public key for decryption.
   (d) Computer B randomly chooses a key for decryption.

A. Elahi, A. Cushman, *Computer Networks*,
https://doi.org/10.1007/978-3-031-42018-4

10. A hash function is used for **digital signature.**

    (a) **Digital signature**
    (b) Digital certificates
    (c) Encryption
    (d) Authentication

Selected Answers: Short Answer

2. List the elements of network security and explain the function of each.

    Confidentiality – information sent through a network must remain confidential to those who are not authorized. This is accomplished through encryption.
    Authentication – The identity of users in a network must be authenticated to ensure that the identity is who they say they are.
    Integrity – Data should be protected against tampering to ensure that messages and information sent arrive at a destination in-tact and it its original state.
    Non-repudiation – A recipient should be provided with proof of origin.

4. What is the definition of cryptography?
    Cryptography is a technique used to establish secure communications through the use of encryption.

6. What is a cipher?
    A cipher is an encryption or decryption algorithm.

8. Explain symmetric cryptography.
    In symmetric cryptography, both sides encrypt and decrypt using the same key, which is kept private from everybody else besides the two communicating.

10. Distinguish between stream ciphers and block ciphers
    Stream ciphers encrypt one byte at a time, while block ciphers encrypt multiple bits in groups at a time called blocks.

12. List three encryption algorithms that use block ciphers.
    DES, 3DES, and AES are all block ciphers.

14. What does AES stand for?
    AES stands for Advanced Encryption Standard.

16. Name an algorithm which generates asymmetric keys.
    RSA generates asymmetric keys.

18. What are the characteristics of a hash value?
    A hash value should be collision resistant (difficult to find two messages that produce the same hash), respond to changes of any bit of the source with a complete change of the resulting hash value, and be impossible to reverse.

20. What is the equation for elliptic curve?
   The elliptic curve equation is $y^2 = x^3 + ax + b$.
22. What are the advantages of ECC?
   ECC has a shorter key length than RSA, has faster key and signature generation, as well as faster encryption and decryption.
24. Explain digital signatures.
   A digital signature is a use of public key cryptography, where a private key may be used to sign a document or hash to establish understanding or acceptance of contents of a digital document.
26. Find the public key, the private key, and n using the RSA (assume the two prime numbers p = 7 and q = 11)

$$N = p^*q$$
$$N = 7^*11 = 77$$

$$Z = (p-1)*(q-1)$$
$$Z = (7-1)*(11-1) = 60$$

$Kp = 13$

13 is less than N (77) and shares no common factors with Z (60)

$$Ks = (1+n^*Z)/Kp$$
$$Ks = (1+n^*60)/13$$
When $n = 8$, $Ks = (1+8^*60)/13 = 37$

   Therefore, the public key is 13, the private key is 37, and N is 77
28. What are the applications of MAC?
   MAC is used for message integrity and authentication of the sender, to determine if a message was tampered while in transit

Chapter 15: Network Security

Selected Answers: Multiple Choice Questions

2. The Secure Socket Layer is located between the **TCP and HTTPS protocols.**

    (a) TCP and IP protocols
    (b) TCP and HTTP protocols
    (c) **TCP and HTTPS protocols**
    (d) TCP and DNS protocols

4. Which of the following is not an SSL/TLS handshake packet? **Server Certificate**

    (a) Client Hello
    (b) Server Hello
    (c) Client Key Exchange
    (d) **Server Certificate**

6. VPNs use **the Internet** to connect networks.

    (a) Leased lines
    (b) Modems
    (c) **The Internet**
    (d) Public Networks

8. Which of the following servers are usually in the DMZ? **DNS Server**

    (a) **DNS Server**
    (b) Database Server
    (c) Accounting Server
    (d) Print Server

10. IPsec is used for **VPNs.**

   (a) Ethernets
   (b) The Internet
   (c) **VPNs**
   (d) VLANs

12. The X.509 is the standard for **certificates.**

   (a) HTTPs
   (b) Telnet
   (c) EAP
   (d) **Certificates**

14. SSH is used to secure **Telnet and FTP**

   (a) Telnet
   (b) FTP
   (c) VLAN
   (d) **(a)** & **(b)**

16. Which of the following firewalls perform NAT? **Application Proxy**

   (a) Packet filtering
   (b) Stateful
   (c) **Application Proxy**
   (d) None of the above

18. What methods does WPA2-Enterprise use for authentication? **RADIUS**

   (a) Pre-shared Keys
   (b) IEEE 802.1X with EAP
   (c) **RADIUS Server**
   (d) None of the above

Selected Answers: Short Answer

2. List the SSL/TLS protocols.
   The SSL/TLS protocols are divided between the Handshake protocols and the Record protocol.
4. List the information in the SSL/TLS Client Hello.
   The SSL/TLS Client Hello contains:

   • **The SSL/TLS version**
   • **A random number**
   • **The Session ID**
   • **A list of usable Cipher Suites**
   • **The compression method (optional)**

6. What is an application of VPNs?
   VPNs are used to transmit information through secure tunnels over the public Internet without the need for privately leased lines, such as to connect remote clients to an on-premises network.
8. List the IPsec modes of operation.
   IPSec consists of Application Header (AH) and Encapsulating Payload (ESP)
10. What are the applications of SSH?
    SSH is a replacement for Telnet and FTP, which allows for secure remote login.
12. What is 802.1x?
    IEEE 802.1x is used for port-based access control and authentication.
14. List three EAP protocols.
    Some EAP protocols include EAP-TLS, LEAP, and PEAP.
16. What is the name of the algorithm used for encryption by WPA2?
    WPA2 uses AES as an encryption algorithm.
18. Show the block diagram for Counter Mode (CTR).

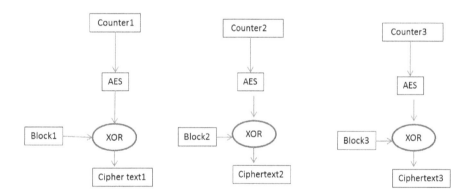

20. What are the different types of firewalls?
    Types of firewall include Packet Filtering Firewalls, NAT Firewalls, and Stateful Firewalls.

Bibliography

Ata Elahi, *Communication Network Technology*, Thomson Delmar Learning, 2001

Ata Elahi, Mehran Elahi, *Data Network, & Internet Communications Technology*, Thomson Delmar Learning. 2006

Beasley, J. *Networking,* Prentice Hall, 2004

Bisaillon, T. Werner, B. *TCP/IP with Windows Illustrated*, McGraw-Hill, 1998

Black, U. ATM: *Foundation for Broadband Net*works, Prentice-Hall, 1995

Black, U. *Frame Relay Networks*, McGraw-Hill, 1996

Comer, E. D. *Computer Networks and Internets*, 6nd edition, Prentice Hall, 2015

Elahi, A. *Computer Systems: Digital Design, Fundamentals of Computer Architecture and ARM Assembly Language*, 2nd edition, Springer, 2022

Elahi, Gschwender, *ZigBee Wireless Sensor and Control Network*, Prentice Hall, 2010

Feit, S. *TCP/IP Architecture and Implementation with IPv6 and IP Security*, 2nd Edition, McGraw-Hill, 1997

FitzGerald, Dennis. *Business Data Communications and Networking*, Wiley, 2020

Garcia, A and Widjaja *Communication Networks,* McGraw-Hill, 2004

Halsall, F. *Multimedia Communications,* Addison-Wesley, 2001

Held, G. Sarch, R. *Data Communications*, 3rd edition, McGraw-Hill, 1995

Hura, G and Singhal, M. *Data and Computer Communications*, CRC 2000

Panko, R. *Business Data Communications and Networking*, 2nd edition, Prentice Hall, 1999

Peterson, Davie. *Computer Networks*, MK, 2020

Shaeda, N.K. *Multimedia Information Networking*, Prentice Hall, 1999

Shay, W. A. *Understanding Data Communications & Networks*, 3Ed edition, Thomson, 2004

Stalling, W. *Data and Computer Communications*, 10nd edition, Prentice-Hall 2016a

Stalling, W. *Network Security Essentials: Applications and Standards,* Prentice-Hall 2016b

Stalling, W. *ISDN and Broadband ISDN with Frame Relay and ATM*, 3rd Edition, Prentice-Hall, 1995

Stalling, W. *Wireless Communications Networks and Systems*, Prentice-Hall, 2015

Stevens, W.R. *TCP/IP Illustrated*, Vol. 1, Addison-Wesley, 2011

Thomas, S.A. *IPng and the TCP/IP Protocols: Implementing the Next Generation Internet*, Wiley, 1996

Wu, C. H. Irvin, J.D. *Emerging Multimedia Computer Communication Technologies*, Prentice-Hall, 1999

A. Elahi, A. Cushman, *Computer Networks*,
https://doi.org/10.1007/978-3-031-42018-4

References by Topic

Advance Metering Infrastructure (AMI)

https://www.energy.gov/sites/prod/files/2016/12/f34/AMI%20Summary%20Report_09-26-16.pdf

ZigBee Specification

https://csa-iot.org/wp-content/uploads/2022/01/docs-05-3474-22-0csg-zigbee-specification-1.pdf

LoRa Wireless

https://www.semtech.com/lora/what-is-lora

6LowPAN

https://www.rfc-editor.org/rfc/rfc8138.html

WirelessHart

https://www.cse.wustl.edu/~lu/cse521s/Slides/wirelesshart.pdf

Sigfox

https://www.sigfox.com/

Protocol Analyzer

https://www.wireshark.org

Cable Modem

https://www.sis.pitt.edu/mbsclass/standards/langer/cablest1.html
http://www.cablelabs.com

Gigabit Ethernet

http://www.gigabit-ethrnet.org

Internet

https://www.ietf.org/rfc/
https://www.internetsociety.org/internet/history-internet/brief-history-internet/
https://www.w3schools.in/types-of-network-protocols-and-their-uses
http://www.dhcp.org/
http://www.internetvalley.com
http://www.broadwatch.com
http://www.vbns.com
http://rs.internic.net

Firewall

https://www.firewall.com
https://www.cisco.com/c/en/us/products/security/firewalls/what-is-a-firewall.html
https://www.cisa.gov/news-events/news/understanding-firewalls-home-and-small-office-use
https://usa.kaspersky.com/resource-center/definitions/firewall

SONET

http://www.atis.org
http://bugs.wpi.edu:8080/EE535/virtext.html
http://www.niuf.nist.gov

Networks and Internet Security

https://www.ibm.com/topics/network-security
http://www.cisco.com/warp/public/cc/neso/sqso/eqso/ipsec_wp.htm
http://wp.netscape.com

http://www.ssh.fi/support/cryptography/introduction/random.html
http://www.iacr.org/
https://www.cisecurity.org/

Voice Over IP

http://www.iec.org
http://www.h323forum.org/papers/
http://voip.internet2.edu

Wireless Networking

https://grouper.ieee.org/groups/802/11/Reports/802.11_Timelines.htm
https://www.wi-fi.org/
http://grouper.ieee.org/groups/802/16/index.html

IEEE Standards

http://standards.ieee.org/getieee802/

Index

© The Editor(s) (if applicable) and The Author(s), under exclusive license to
Springer Nature Switzerland AG 2024
A. Elahi, A. Cushman, *Computer Networks*,
https://doi.org/10.1007/978-3-031-42018-4

Printed in the USA
CPSIA information can be obtained
at www.ICGtesting.com
LVHW021211211023
761742LV00005B/34